Gerd Habenicht
Applied Adhesive Bonding

Further Reading

W. Brockmann, P. L. Geiß, J. Klingen, B. Schröder

Adhesive Bonding
Materials, Applications and Technology

2009
ISBN: 978-3-527-31898-8

Gerd Habenicht

Applied Adhesive Bonding

A Practical Guide for Flawless Results

Translated by Christine Ahner

WILEY-VCH Verlag GmbH & Co. KGaA

Author

Prof. Dr. rer. nat. Gerd Habenicht
Seestrasse 33
82237 Wörthsee
Germany

Translator

Christine Ahner
translate economy
Frhr.-v.-Eichendorff-Strasse 8/1
88239 Wangen im Allgäu
Germany

Originally published in German language by Vieweg+Teubner as "Gerd Habenicht: Kleben – erfolgreich und fehlerfrei", 4th edition, 2006.
© Vieweg+Teubner Verlag/GWV Fachverlage GmbH, Wiesbaden, 2006

■ All books published by Wiley-VCH are carefully produced. Nevertheless, authors, editors, and publisher do not warrant the information contained in these books, including this book, to be free of errors. Readers are advised to keep in mind that statements, data, illustrations, procedural details or other items may inadvertently be inaccurate.

Library of Congress Card No.: applied for

British Library Cataloguing-in-Publication Data
A catalogue record for this book is available from the British Library.

Bibliographic information published by the Deutsche Nationalbibliothek
The Deutsche Nationalbibliothek lists this publication in the Deutsche Nationalbibliografie; detailed bibliographic data are available on the Internet at http://dnb.d-nb.de

© 2009 WILEY-VCH Verlag GmbH & Co. KGaA, Weinheim

All rights reserved (including those of translation into other languages). No part of this book may be reproduced in any form – by photoprinting, microfilm, or any other means – nor transmitted or translated into a machine language without written permission from the publishers. Registered names, trademarks, etc. used in this book, even when not specifically marked as such, are not to be considered unprotected by law.

Printed in the Federal Republic of Germany
Printed on acid-free paper

Typesetting   Manuela Treindl, Laaber
Printing   Strauss GmbH, Mörlenbach
Bookbinding   Litges & Dopf Buchbinderei GmbH, Heppenheim

ISBN: 978-3-527-32014-1

Contents

Preface *XI*

1 Introduction *1*
1.1 Bonding as a Joining Process *1*
1.2 Advantages and Disadvantages of Bonding *1*
1.3 Terms and Definitions *3*

2 Structure and Classification of Adhesives *5*
2.1 Structure of Adhesives *5*
2.1.1 Carbon as Central Element *5*
2.1.2 Monomer – Polymer *6*
2.1.3 Polymer Formation *7*
2.2 Classification of Adhesives *8*
2.2.1 Adhesives Curing by Chemical Reaction (Reactive Adhesives) *8*
2.2.2 Adhesives Curing without Chemical Reaction (Physically Setting Adhesives) *8*
2.2.3 Solvent-Containing and Solvent-Free Adhesives *9*
2.2.4 Adhesives on Natural and Synthetic Basis *10*
2.2.5 Adhesives on Organic and Inorganic Basis *10*
2.2.6 Application-Related Names of Adhesives *11*

3 From Adhesive to Adhesive Layer *13*
3.1 Reactive Adhesives – Fundamentals *13*
3.1.1 Pot Life *14*
3.1.2 Mixing Ratio of the Components *14*
3.1.3 Impact of Time on Adhesive Curing *15*
3.1.4 Impact of Temperature on Adhesive Curing *16*
3.2 Two-Component and One-Component Reactive Adhesives *17*
3.2.1 Two-Component Reactive Adhesives *18*
3.2.2 One-Component Reactive Adhesives *18*
3.3 Properties of Adhesive Layers *19*
3.3.1 Thermoplastics *19*

Applied Adhesive Bonding: A Practical Guide for Flawless Results. Gerd Habenicht
Copyright © 2009 WILEY-VCH Verlag GmbH & Co. KGaA, Weinheim
ISBN: 978-3-527-32014-1

3.3.2	Thermoset Plastics	20
3.3.3	Elastomers	21
3.3.4	Glass Transition Temperature	21
3.3.5	Creep	22

4 Important Reactive Adhesives 23

4.1	Epoxy Resin Adhesives	23
4.1.1	Two Component Epoxy Resin Adhesives	23
4.1.2	One-Component Epoxy Resin Adhesives	25
4.1.3	Reactive Epoxy Resin Hot-Melt Adhesives	25
4.1.4	Properties and Application of Epoxy Resin Adhesives	26
4.2	Polyurethane (PUR) Adhesives	26
4.2.1	Two-Component Polyurethane Adhesives (Solvent-Free)	27
4.2.2	One-Component Polyurethane Adhesives (Solvent-Free)	27
4.2.3	Reactive Polyurethane Hot-Melt Adhesives (Solvent-Free)	29
4.2.4	One-Component Polyurethane Solvent-Based Adhesives	30
4.2.5	Two-Component Polyurethane Solvent-Based Adhesives	30
4.2.6	Polyurethane Dispersion Adhesives	30
4.3	Acrylic Adhesives	31
4.3.1	Cyanoacrylate Adhesives	33
4.3.2	Radiation-Curing Adhesives	34
4.3.3	Methacrylate Adhesives	35
4.3.4	Anaerobic Adhesives	37
4.4	Unsaturated Polyester Resins (UP-Resins)	39
4.5	Phenolic Adhesives	39
4.6	Silicones	40
4.7	Summary Reactive Adhesives	41
4.8	Film Adhesives	42
4.9	Sealing Materials	42
4.10	Polymer Mortars	43

5 Physically Setting Adhesives 45

5.1	Hot-Melt Adhesives	45
5.2	Solvent-Based Adhesives	47
5.3	Contact Adhesives	50
5.4	Dispersion Adhesives	51
5.5	Plastisols	53
5.6	Pressure-Sensitive Adhesives, Adhesive Tapes	53
5.7	Adhesive Strips	55
5.8	Glue Sticks	55
5.9	Adhesives Based on Natural Raw Materials	55
5.10	Adhesives on an Inorganic Basis	56

6	**Adhesive Forces in Bonded Joints**	57
6.1	Adhesive Forces Between Adhesive Layer and Adherend (Adhesion)	57
6.2	Wetting	59
6.3	Surface Tension	60
6.4	Adhesive Forces Inside an Adhesive Layer (Cohesion)	61

7	**Production of Bonded Joints**	63
7.1	Surface Treatment	64
7.1.1	Surface Preparation	64
7.1.1.1	Cleaning	64
7.1.1.2	Adjusting	64
7.1.1.3	Degreasing	64
7.1.1.4	Degreasing Agents	65
7.1.2	Surface Pretreatment	66
7.1.2.1	Mechanical Surface Pretreatment	67
7.1.2.2	Physical and Chemical Surface Pretreatment	68
7.1.2.3	Pickling	69
7.1.2.4	Surface Layers and Creep Corrosion	69
7.1.3	Surface Post-Treatment	70
7.1.3.1	Primer	70
7.1.3.2	Climatization	71
7.2	Adhesive Processing	71
7.2.1	Adhesive Preparation	71
7.2.1.1	Viscosity Adjustment	71
7.2.1.2	Homogenization	71
7.2.1.3	Climatization	72
7.2.2	Adhesive Mixing	72
7.2.2.1	Industrial Processing	72
7.2.2.2	Application in Workshops	72
7.2.2.3	Dynamic Mixers	73
7.2.2.4	Static Mixers	74
7.2.3	Adhesive Application	75
7.2.3.1	Application Methods	76
7.2.3.2	Laminating	76
7.2.3.3	Amount Applied	77
7.2.4	Fixing of Adherends	78
7.2.5	Adhesive Curing	79
7.2.5.1	Drying, Evaporating	80
7.2.5.2	Curing	80
7.3	Repair Bonding	81
7.3.1	Metal Components	81
7.3.2	Plastics	83
7.3.2.1	Rigid Materials	83
7.3.2.2	PVC Films	84

7.3.2.3 Gummed Fiber Fabric   84
7.4   Mistake Possibilities in Bonding and Remedial Actions   85
7.5   Safety Measures in Adhesive Processing   88
7.5.1   Workplace Prerequisites for Adhesive Processing   88
7.5.2   Rules of Conduct in Adhesive Processing   89
7.6   Quality Assurance   90
7.7   Adhesive-Bonding Training   91

8   Adhesive Selection   93
8.1   Preliminary Notes   93
8.2   Determining Factors for the Selection of Adhesives   94
8.2.1   Adherend Properties   95
8.2.2   Demands on Bonded Joints   96
8.2.3   Preconditions in Manufacturing   96
8.2.4   Processing Parameters of Adhesives   97
8.2.5   Property-Related Parameters of Adhesives and Adhesive Layers   97
8.2.5.1   One-Component Reactive Adhesives   98
8.2.5.2   Two-Component Reactive Adhesives   99
8.2.5.3   Physically Setting Adhesives   99
8.3   Selection Criteria   101

9   Adhesive Properties of Important Materials   105
9.1   Metals   105
9.1.1   Fundamentals   105
9.1.1.1   Strength   105
9.1.1.2   Impermeability Towards Solvents   106
9.1.1.3   Insolubility in Solvents   106
9.1.1.4   Thermal Conductivity   106
9.1.1.5   Temperature Resistance   106
9.1.2   Surface Pretreatment   106
9.1.3   Bondability of Important Metals   107
9.1.3.1   Aluminum and Al-Alloys   107
9.1.3.2   Noble Metals   107
9.1.3.3   Stainless Steels   107
9.1.3.4   Copper   107
9.1.3.5   Brass   108
9.1.3.6   Steels, General Constructional Steels   108
9.1.3.7   Galvanized Steels, Zinc   108
9.1.4   Adhesives for Bonded Metal Joints   108
9.2   Plastics   109
9.2.1   Fundamentals   109
9.2.2   Classification of Plastics   110
9.2.3   Identification of Plastics   112
9.2.4   Surface Pretreatment   112
9.2.4.1   Corona Method   112

9.2.4.2	Low-Pressure Plasma	*113*
9.2.4.3	Atmospheric Pressure Plasma	*113*
9.2.4.4	Flame Treatment (Kreidl Method)	*113*
9.2.4.5	Mechanical Methods	*114*
9.2.5	Plastics Soluble or Swellable in Organic Solvents	*114*
9.2.6	Plastics Insoluble or not Swellable in Organic Solvents	*116*
9.2.7	Plastic Foams	*116*
9.2.8	Bonding of Plastics to Metals	*117*
9.2.9	Bonding of Plasticizer-Containing Plastics	*118*
9.3	Glass	*118*
9.3.1	Surface Pretreatment	*118*
9.3.2	Glass–Glass-Bonded Joints	*119*
9.3.3	Bonded Glass Joints with Radiation-Curing Adhesives	*119*
9.3.4	Glass–Metal Joints	*120*
9.4	Rubber Products	*121*
9.5	Wood and Wood Products	*122*
9.6	Porous Materials	*123*
10	**Strength, Calculation and Testing of Bonded Joints**	*125*
10.1	The Term "Strength"	*125*
10.2	Test Methods	*128*
10.2.1	Adhesive Strength Testing	*128*
10.2.2	Tensions in Single-Lap Bonded Joints	*129*
10.2.3	Shear Strength Testing	*131*
10.2.4	Peel Resistance Testing	*131*
10.2.5	Test Methods for Short-Term and Long-Term Stresses	*133*
10.3	Elastic Bonding	*135*
10.4	Shaft-to-Hub Joints	*137*
11	**Constructive Design of Bonded Joints**	*139*
12	**References**	*145*
13	**A Selection of Common Terms in Bonding Technology**	*149*
	Index	*167*

Preface

Bonding especially concerning this book "adhesive bonding", is, without doubt, the oldest of the three joining procedures welding, soldering and bonding, as can be derived from depictions in millennia-old mural paintings. The procedure of adhesive sealing, very closely related to bonding has also been known for ages and had already been applied to the construction of Noah's ark for the sealing of the wooden ship walls with pitch, as narrated in the Bible (First Book of Mose, Chapter 6, verse 14).

Moreover, it has to be mentioned that bonding is the only procedure of the three mentioned that also occurs in nature. For the construction of their honeycombs, for example, bees use their endogenous adhesive secretions together with finest wood fibers to create firmly bonded constructions; swallows affix their nests to masonry; the sundew uses adhesive substances on its tentacles to catch insects as food; and finally even the construction of a spider net is based on the principle of bonding. Man also produces an "adhesive" as a blood component, the fibrinogen, which converts into fibrin through chemical reaction, thus "bonding" tissue components of a wound.

Despite the increasing development of application possibilities for bonding in the remote and recent past, knowledge of adhesives and their processing is still limited and hardly anyone takes into consideration that bonding is a manufacturing method that has to be learned – if it shall be successful, although it seems to be so easy (as it is often represented) that no special knowledge is required to firmly join two materials after the application of the adhesive and the compression of the adherends. Then, if the joint does not hold, it is the fault of the adhesive and the results give rise to doubts as to the reliability of this bonding procedure. However, that such doubts are unfounded is proven by a great number of successful applications with high operational demands, for example, on bonded rotor blades of helicopters or glass panes bonded in the bodywork of motorcars.

This book is intended to provide both basic knowledge in the successful application of bonding in industry, trade or in the private sector and the required information for a professional training in bonding techniques. Especially in the industrial sector, this is a prerequisite indispensable for a quality-oriented mastering of the production system "bonding". The author has addressed himself to the task of describing the essential facts vital for the successful application of bonding in a

Applied Adhesive Bonding: A Practical Guide for Flawless Results. Gerd Habenicht
Copyright © 2009 WILEY-VCH Verlag GmbH & Co. KGaA, Weinheim
ISBN: 978-3-527-32014-1

way comprehensible even for nonscientists. This includes explanations of the chemistry of adhesives, the kind of bonding strengths, tips regarding the making of bonded joints, the description of failure possibilities and their avoidance as well as safety measures, tests and structuring principles. For the acquisition of the learning matter it is recommended to work through the book from the beginning instead of randomly reading individual chapters, since particularly with regard to the technical terms and their explanation, a continuing structure was chosen.

May this book reach its aim and impart the knowledge of adhesive-bonding and contribute to the successful application of this joining process.

Wörthsee/Steinebach, December 2008 *Gerd Habenicht*

1
Introduction

1.1
Bonding as a Joining Process

Adhesive-bonding is assigned to the *materially joined* processes. Bonding processes serve the production of joints of materials of the same kind or of material combinations. The term "materially joined process", which also includes welding and soldering, derives from the fact that the bond occurs by a separately added material, that is

- the adhesive in the case of bonding,
- the welding additive material in the case of welding, and
- solder in the case of soldering.

In addition, there are

- *positive* joints, for example, folding, indented joining;
- *nonpositive* joints, for example, pressing, clamping, screwing, riveting (Figure 1.1).

1.2
Advantages and Disadvantages of Bonding

Compared to some of the joining processes depicted in Figure 1.1, bonding shows remarkable *advantages*:

- The adherends are not weakened by bores as it is the case, for example, when screwing and riveting. Thus power transmission is surface-related instead of spot-related (Figure 1.2).
- There adherends are not stressed by high temperatures, as in welding and, partly, even in soldering. Thus, thermally caused modifications of material properties are prevented, which enables heat-sensitive materials to be joined.
- Adhesive-bonding allows extremely different materials to be joined with themselves or with other materials while retaining their specific characteristics. In the latter case, it is possible to utilize the different advantageous properties for innovative composite structures.

Applied Adhesive Bonding: A Practical Guide for Flawless Results. Gerd Habenicht
Copyright © 2009 WILEY-VCH Verlag GmbH & Co. KGaA, Weinheim
ISBN: 978-3-527-32014-1

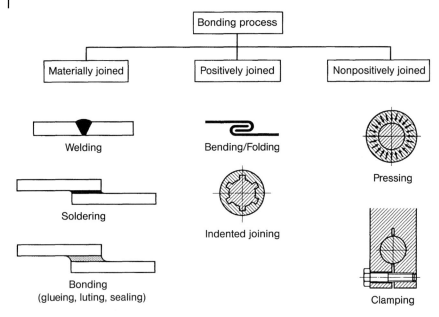

Figure 1.1 Classification of joining processes.

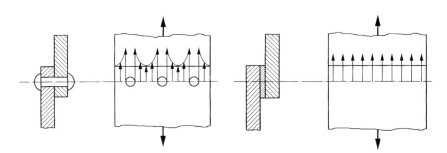

Figure 1.2 Power transmission in riveted (screwed) and bonded joints.

- Bonding as a joining process enables joints of very thin materials (< 500 µm). This procedure is particularly important for the manufacturing of lightweight constructions and the related weight reduction (*inter alia* in aerospace manufacturing). Furthermore, it is the basis of an extremely varied design of film-type laminates in the packaging industry.
- The combination of positive and nonpositive joining processes is important for the optimization of strength, stiffness and corrosion resistance, as the case may be (e.g., folding – bonding in car manufacturing), tightness (e.g., in screw, rivet and spot-welding constructions, shaft-to-collar connections and folds) (Figure 1.1 and 11.5).
- Compared to riveted or screwed connections homogeneous stress distribution when stress-loading (Figure 1.2).

However, these advantages are curtailed by the following *disadvantages:*

- The heat resistance of the adhesive layer is limited. Depending on the basic material of the adhesive, temperatures for continuous stress range between approximately 120 and 300 °C.
- Adhesive layers and their boundary layers towards the adherends' surfaces may be damaged by environmental impacts, such as humidity, which results in a reduction of strength.
- With a few exceptions (e.g., body-in-white manufacturing), the production of bonded joints requires surface treatment of the adherends as an additional production stage.
- In the production of bonded joints, the time required for the relevant reaction kinetics of curing has to be taken into consideration.
- The growing demand for recyclability of industrial products calls for respective design-engineering measures.
- The availability of nondestructive test methods is rather limited.

The essential difference between welding and soldering on the one hand, and bonding on the other hand is the structure of the additive material. Welding additives and solders consist of metals and metal alloys, respectively, which liquefy to a melted mass under the influence of heat (welding torch, soldering iron) and result in a joint after cooling down while integrating parts of the adherends. Adhesives, in comparison, consist of chemical compounds and structures on a basis different from that of metals. These relations are described in Chapter 2.

1.3
Terms and Definitions

Binding terms are a prerequisite to ensure quality-determining production flows in industrial processes. The following terms apply to the manufacturing system of "bonding":

1. Bonding: Joining of same or different materials under the application of adhesives.
2. Adhesive: Nonmetal, liquid, paste-like or even solid material, joining adherends by means of adhesion forces (surface adhesion) and cohesion forces (inner stability of the adhesive layer) (Chapter 6).
3. Adhesive layer: Adhesive layer between the adherends, set (cured) or still not set.
4. Boundary layer: Zone between adherends surface and adhesive layer where adhesion and bonding strengths are effective.
5. Glueline: Space between two adherend surfaces filled with an adhesive layer.
6. Adherend surface: The glued surface or surface to be glued of an adherend or a bonded joint.
7. Bonded joint: Joint of adherends, obtained by an adhesive.

1 Introduction

8. Adherend: Body bonded or to be bonded to another body.
9. Setting, curing: Solidification of the liquid adhesive layer.
10. Structural bonding: Structural design with high strength and stiffness resp., with regular and favorable stress distribution (contrary: fixing bonding, e.g., in case of wallpaper) possible through bonding.

Figure 1.3 shows the structure of a single-lap bonded joint with the most important terms.

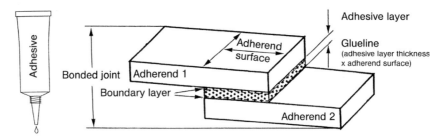

Figure 1.3 Adhesive terms.

Complementary literature to Chapter 1 – general overviews:
[A2, B5, C2, K1, L1, P3].

2
Structure and Classification of Adhesives

2.1
Structure of Adhesives

2.1.1
Carbon as Central Element

With regard to their chemical structure, adhesives are to be assigned to the *organic* compounds. In contrast to *inorganic* chemistry, which treats matters of the inanimate nature (e.g., minerals, metals), organic chemistry deals with compounds of carbon as the central element of diverse matters that make up the animate nature (e.g., plant and animal products such as wood, proteins, resins, fats, petroleum).

The special feature of carbon, and thus its dominating position among all known elements, is the fact that an almost unlimited number of compounds with carbon itself and with a multitude of other elements is possible. Each carbon atom (atoms are the smallest "components" characteristic for the properties of an element) shows four "arms" it can "spread" to form a bond. In chemistry, these "arms" are depicted by simple lines and, derived from the Latin word *valentia* = power, strength, they are called *valencies*:

$$-\underset{|}{\overset{|}{C}}-$$

These valencies or bonding possibilities between individual carbon atoms result in long chains,

$$-\underset{|}{\overset{|}{C}}-\underset{|}{\overset{|}{C}}-\underset{|}{\overset{|}{C}}-\underset{|}{\overset{|}{C}}-\underset{|}{\overset{|}{C}}-\underset{|}{\overset{|}{C}}-\underset{|}{\overset{|}{C}}-$$

that may also possess branchings, crosslinked or annular structures:

Applied Adhesive Bonding: A Practical Guide for Flawless Results. Gerd Habenicht
Copyright © 2009 WILEY-VCH Verlag GmbH & Co. KGaA, Weinheim
ISBN: 978-3-527-32014-1

2 Structure and Classification of Adhesives

$$\begin{array}{c} |\\ -C-\\ |\\ -C-\\ |\\ |\quad|\quad|\quad|\quad|\quad|\quad|\quad|\\ -C-C-C-C-C-C-C-C-\\ |\quad|\quad|\quad|\quad|\quad|\quad|\quad|\\ -C-\\ |\\ -C-\\ |\\ -C-\\ | \end{array}$$

The formation of two bonds between two carbon atoms is possible as well,

$$-\underset{|}{C}=\underset{|}{C}-$$

furthermore, bonds with other elements exist, as, for example, with

hydrogen $\quad H-\underset{|}{\overset{|}{C}}-$

oxygen $\quad O=\underset{|}{\overset{}{C}}$

The individual elements have different numbers of valencies and thus of bonding possibilities, which are predetermined by the structure of their atoms. From these explanations it can be derived that there is a multitude of different organic compounds (more than 1 million) in which above all the elements

	chemical symbols	
carbon	C	(from the Latin word carbo),
hydrogen	H	(from the Latin word hydrogenium),
oxygen	O	(from the Latin word oxigenium),
nitrogen	N	(from the Latin word nitrogenium)

are involved. These organic compounds also comprise by far the largest part of the adhesives. Since in their structure, they are again very similar to the plastics known to us, partly even identical, they are also assigned to the products of the "plastic age". The modern "synthetic" adhesives became known only about 100 years ago. The first synthetic material with technical significance was "Bakelite", named after its inventor, the Belgian L. H. Baekeland (1863–1944), a phenol-formaldehyde resin nowadays still applied as synthetic material.

2.1.2
Monomer – Polymer

To further describe adhesives, the explanation of two important technical terms is required (Figure 2.1).

- *Monomer:* This term derives from the Greek language (monos = separate, individual) and indicates the single "components" that combine to polymers due to a chemical reaction.

- *Polymer:* Also of Greek origin (polys = many, meros = portion, part), meaning something like a system of "many parts".

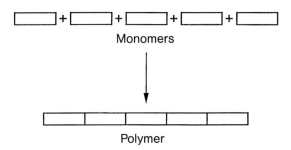

Figure 2.1 Polymer formation from monomers (I).

2.1.3
Polymer Formation

The depiction of the polymer formation can be compared with the arrangement of railway wagons. Due to the "hooks and eyes" at the wagons, any number of wagons (monomers) can connect to a train (polymer) (Figure 2.2).

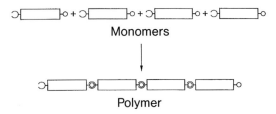

Figure 2.2 Polymer formation from monomers (II).

For this purpose, monomers show a special combination of different elements, so-called "reactive groups" that, instead of the metaphorically called "hooks and eyes", combine with the groups of neighboring monomers by chemical reaction. Thus, "polymer structures" develop from straight and branched or crosslinked chains. These reactive groups will be explained in more detail when the most important adhesives are discussed.

If only a limited number of monomers combine by chemical reaction, then one talks of *prepolymers*, a preliminary stage of polymers, however, still showing reactive groups. They are partly applied in mixtures with similarly structured monomers. To simplify matters, the term monomer continues to be used in this book.

2.2
Classification of Adhesives

The various structural possibilities of adhesives and the related variants of their processing amount to thousands of different "compositions" or "formulations", which are applied in practice. This makes it difficult for the user to find the adequate adhesive, and thus it is often asked which adhesive is suitable for a certain material. The following description of the properties will facilitate the answer.

2.2.1
Adhesives Curing by Chemical Reaction (Reactive Adhesives)

In this case, the liquid adhesive applied to the adherends consists of the monomer molecules ready for a chemical reaction (Sections 2.1.2 and 2.1.3). Due to their "small size" they are mostly liquid. After the application of the adhesive and the joining of the adherends to be bonded, a chemical reaction occurs in the glueline. From the (liquid) monomers the solid ("hard") adhesive layer develops. This time-dependent process is called *curing* or *setting*. Since it is triggered by a chemical reaction, one talks of *chemically reacting adhesives* or of *reactive adhesives*.

2.2.2
Adhesives Curing without Chemical Reaction (Physically Setting Adhesives)

The process of polymer formation by monomers reacting with each other (Section 2.1.3) can already be carried out by the adhesive manufacturer, that is, before the user applies the adhesive. However, the consequence will be that the existing polymers, due to their long chains or even branched net structures, are no longer liquid and cannot be processed in this form. To enable their application they have to be suitably transferred into a liquid state. There are different possibilities for this "liquefaction":

- The polymers are dissolved in organic solvents. Such adhesives are called *solvent-based adhesives* (Section 2.2.3).

- Also water may serve as a liquid medium, in which the finely distributed polymeres are "floating". These adhesives are commercially available as *dispersions* (Latin *dispergere* = finely distribute) (Section 5.4).

- There are also polymers that can be brought to melting by heat supply. They are applied to the adherends in molten and solvent-free form. When the adhesive melt has cooled down, a bonded joint develops. Such adhesives, processed by melting and cooling down, are called *hot-melt adhesives* (Section 5.1).

- Furthermore, polymer layers applied to the respective substrates are known that show their own adhesiveness due to the addition of tackifying components (e.g., resins). With the application of adequate surface pressure, these systems known as *pressure-sensitive adhesives* (Section 5.6) result in a bonded joint.

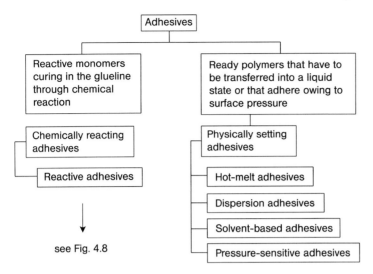

Figure 2.3 Classification of adhesives according to their way of curing.

If solvent-based adhesives or even dispersions are applied to the adherends, the solvents or the water will have to escape from the liquid adhesive layer before the adherends will be joined, they have to "evaporate". A chemical reaction does not take place; the process of solvent evaporation is a so-called "physical process". Thus, instead of chemically reacting or reactive adhesives, one talks of *physically setting adhesives*. Since the cooling of a melt, in the case of hot-melt adhesives, or pressure application, in the case of the pressure-sensitive adhesives, are also physical processes, they, too, belong to this group. In contrast to the curing of reactive adhesives in case of physically setting adhesives, this process, as the name already implies, is called "setting", respectively, "solidification". These relationships are schematically illustrated in Figure 2.3.

In addition to the classification criteria described in the two preceding sections further characteristics are quite common (Sections 2.2.3 to 2.2.6).

2.2.3
Solvent-Containing and Solvent-Free Adhesives

As described above, adhesives on polymer basis have to be transferred into a processible state by using adequate solvents or water. Thus, the important group of *solvent-based adhesives* develops, in contrast to *reactive adhesives* processed in the form of monomers, which, due to their mostly liquid or pastelike state, do not require solvents. In the common language use, only products containing organic and, in the majority of cases, combustible solvents in their formulations are called solvent-based adhesives.

2.2.4
Adhesives on Natural and Synthetic Basis

Another possibility for adhesives to be classified is the differentiation between organic compounds of natural products, so-called "natural" adhesives, and products resulting from targeted chemical reactions, so-called "synthetic" adhesives. Many substances are known from daily life, which show a natural tack, for example, tree resins, plant juices, waxes, proteins, gelatine, casein, starch. In comparison to synthetically produced adhesives they are heavily lagging behind in terms of quantity, however, they partly show excellent properties in special applications, such as casein adhesives for bottle labelling.

2.2.5
Adhesives on Organic and Inorganic Basis

As mentioned in Section 2.1.1, chemistry is subdivided into "organic" and "inorganic" chemistry. Therefore apart from adhesives on organic basis, even those on inorganic bases are applied. Due to their chemical structure, their advantage is mainly to be seen in the excellent long-term resistance of their adhesive layers against heat at temperatures of up to 500 °C, in special cases even more. Important fields of application are the glass/socket bonding or the glueing in place of energy supply wires in the filament and halogen bulb manufacturing.

In Figure 2.4, the adhesives described in the previous Sections 2.2.4 and 2.2.5 are classified by their chemical basis.

Note: Due to their chemical structure, silicones represent compounds with both organic and inorganic structural characteristics.

Another kind of adhesive classification will be given in Section 3.3, Figure 3.7.

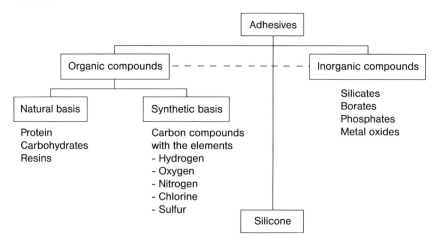

Figure 2.4 Adhesives classified by their chemical basis.

2.2.6
Application-Related Names of Adhesives

Adhesive names referring to special ways of applications are also very common, such as

- Cold glue or hot curing adhesive as a reference to the *processing temperature*.
- Pressure-sensitive adhesive, hot-melt adhesive, contact adhesive, two-component adhesive as a reference to a certain *processing procedure*.
- Wood glue, wallpaper paste, labelling adhesive, metal adhesive as reference to a certain *purpose of use*.
- Film adhesive, glue powder, solvent-based adhesive as a reference to the *form of supply*.
- Epoxy resin adhesive, methacrylate adhesive, polyurethane adhesive as reference to the applied *adhesive basis (adhesive basic material)*.

Glues or pastes are adhesives with relatively high viscosity based on animal and/or plant elements (probably mixed with synthetic elements) using water as solvent, respectively, expanding agent.

The term *all-purpose adhesive* requires a more critical examination. This term often suggests a "universal adhesive" suitable for any application and any stress of most different materials. The reader will understand why such adhesives are suitable for special applications, however, not for *all* purposes, in particular after having studied the explanations regarding the selection of adhesives in Chapter 8.

Thus, *adhesives* are products that – according to their chemical composition and the given physical state – are applied to the adherends where they develop an adhesive layer either by chemical reaction or by physical setting.

Nowadays, adhesives are regarded as important materials among the resources for bonding as a process of undoubtedly high technical standard.

Complementary literature to Chapter 2:
[B4, F1, G1].

3
From Adhesive to Adhesive Layer

3.1
Reactive Adhesives – Fundamentals

As explained in Chapter 2, reactive adhesives consist of monomers, respectively, prepolymers with the necessary prerequisites for a chemical reaction. Such prerequisites are their "reactive groups" the molecules are equipped with. They only need the right "impulse" to "trigger off" the reaction. Such an impulse may occur, for example, if a monomer with a "B-eye" is admixed to a monomer with the suitable "A-hook". Then the monomers A and B begin to amalgamate, they "react". The result is a "reactive" mixture with an increasing number of A and B monomers linking to polymer AB (Figure 3.1, simplified depiction, even branched or crosslinked structures develop). In adhesive processing one talks of the two *components* A and B, which, since they usually occur in the liquid state, are mixed according to the specifications in Section 7.2.2.

For reasons of clarity, the following issues have to be explained:

- Pot life.
- Mixing ratio of the components.
- Impact of time on adhesive curing.
- Impact of temperature on adhesive curing.

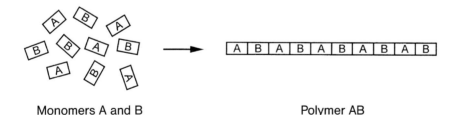

Monomers A and B   Polymer AB

Figure 3.1 Formation of a polymer AB from the components A and B.

Applied Adhesive Bonding: A Practical Guide for Flawless Results. Gerd Habenicht
Copyright © 2009 WILEY-VCH Verlag GmbH & Co. KGaA, Weinheim
ISBN: 978-3-527-32014-1

3.1.1
Pot Life

Since the chemical reaction of the two components A and B begins in a "pot" immediately after mixing, this ready-made adhesive mixture requires speedy application. Otherwise the reaction for the formation of the AB polymer (the adhesive layer) will have developed to such an extent already prior to the application to the adherends that the expected strength of the bonded joint is impaired. Between the mixing of the adhesive mix and its application to the adherends and their fixing only a certain time span is allowed – which may vary for the individual reactive adhesives. This time is called the *pot life*. Depending on the reactivity of the A and B monomers, pot life can lie in the range of minutes or even hours.

The pot life is indicated in the processing instructions of the adhesives; it is, however, subject to certain fluctuations depending on the batch to be mixed. One reason for this is the fact that heat, "reaction heat", develops during the chemical reaction of the components. Since the adhesive mixtures show relatively low thermal conductivity, which means the accruing heat can only slowly dissipate to the environment, it is logical that large batches (e.g., in the range of kilograms) heat up more than small batches in the gram range. Since, in case of higher temperature, the reaction rate of the components A and B is higher than at lower temperature, large batches have a shorter pot life. Thus pot life depends on

- the "reactivity", that is, the rate at which the monomers mutually react,
- the ambient temperature as well as,
- the batch quantity.

The time- and temperature-dependent "gel point" of a reactive adhesive represents the state at which the adhesive transfers from the state of increasingly higher viscosity to the solid state until it reaches its final strength.

3.1.2
Mixing Ratio of the Components

Reactive adhesives are usually available in two different tubes or tins (mostly called *resin* and *hardener*), and have to be mixed according to the weight and volume units defined by the manufacturer. Such adhesives are typically *two-component reactive adhesives* because of the two components A and B that have to be mixed.

But why is the observance of the mixing ratio so important for the adhesives described above? As schematically shown in Figure 3.1, in the case of epoxy resin adhesives (Section 4.1), for example, each monomer A needs a monomer B to form the polymer AB. If compared to B, there is too much of the component A, an excess quantity of A remains, which cannot take part in further reactions. The adhesive layer does not cure completely, thus the bond between the adherends will not be sufficiently strong. Taking Figure 3.2 as an example, it becomes clear that the optimum strength is reached at a mixing ratio of A : B = 1 : 1.

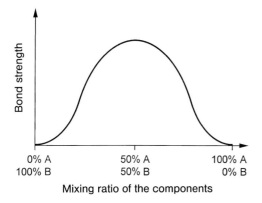

Figure 3.2 Function of the dependence of the bond strength on the mixing ratio of the components.

3.1.3
Impact of Time on Adhesive Curing

We have already learned that the formation of an adhesive layer follows certain chemical reactions. Such reactions are subject to two important factors, that is

- time and
- temperature.

How to explain the influence of time? Let us follow the time-related course of curing in an adhesive mixture of the components A and B, as schematically shown in Figure 3.3:

- At time point 0, the mixture contains only the monomer molecules A and B in the defined ratio.
- Already shortly afterwards (t_1) they both begin to react with each other, the first polymer molecules AB develop.

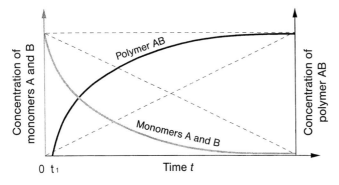

Figure 3.3 Time-related course of polymer formation from monomers.

- Thus with progressing time, the concentration of the monomers A and B decreases, while at the same time the portion of developed polymer molecules AB increases.
- The larger the quantity of the formed polymer AB becomes in the course of time, the lower the concentration of A and B monomers will be. In the end, this leads to the fact that they hardly "meet", since there is too much polymer AB between them.
- At a point in time, which depends on the respective adhesive, the reaction will be finished because of the continuously decreasing portions of A and B that can no longer "find" each other; the adhesive mixture has cured and developed to an adhesive layer. It is in the nature of such chemical reactions that the decrease in A and B and the increase in AB do not occur according to the dotted line, which means straight (linear) as shown in Figure 3.3, but according to the drawn curves. If the reaction rate of A + B → AB is high at the beginning, it will reduce within the course of time. Theoretically, an infinitely long time is required until the reaction is 100% completed. This time depends, *inter alia*, on the size of the formed polymer molecules. These considerations, however, are of little importance for practical application; but they are important in view of the next section discussing the impact of temperature on these reactions.

3.1.4
Impact of Temperature on Adhesive Curing

What can be done to reduce reaction time? Here, temperature is suitable as a second parameter. Chemical procedures can generally be accelerated by heat supply, that is, increase in temperature. The reason is to be found in the increased mobility of molecules occurring with rising temperature. This influence can be explained by the example of water, although here a physical process is concerned and not a chemical one. With temperatures below 0 °C, water is in the solid state, as ice. The water molecules are embedded in a crystal lattice, thus they cannot move. When the ice is heated to above 0 °C, it liquefies to water; the water molecules are no longer fixed in a lattice and can move uncontrolled. This mobility increases with rising temperature until, at 100 °C and at normal pressure, it is sufficient to cause the molecules to give up their liquid bond and to pass to the adjacent air space as water molecules with very high mobility.

Similar reasoning is applicable to monomer molecules. Owing to the increase in temperature their mobility increases and thus the probability to "meet" each other and form the polymer AB; their reaction rate increases, the adhesive cures completely and more quickly. So the curing time of a reactive adhesive can be reduced by heat supply.

This appreciated impact of temperature also enables the production of reactive adhesives for which pot life, usually undesired in the adhesive application, can almost be avoided. For this purpose, monomers are chosen which, due to their chemical "inertness", are not inclined to react with each other at room temperature or below. Thus, in a mixed state, they are "nonreactive" and can be applied

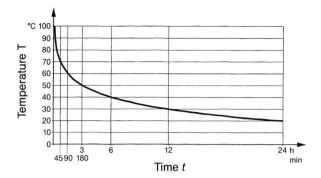

Figure 3.4 Relation between reaction rate and curing temperature in a reactive adhesive.

without pot-life limitation (blocking). Only when the adhesive-coated adherends are heated, does the curing reaction set in Section 3.2.2. The temperatures required vary depending on the monomer structure. In ranges between approx. 60–150 °C one talks of hot-curing adhesives, beyond these temperatures of heat-curing adhesives. Figure 3.4 shows these relations.

From this diagrammed example a time reduction of approx. 20 minutes at 80 °C is deducible for the given adhesive at room temperature with a curing time of 24 hours.

The respective optimum curing times and temperatures are indicated on the technical data sheets or packaging. If automatic mixing and dosing devices are not available for the application of adhesives with a short pot life, it should be ensured that only such quantity of adhesive is mixed that is processable during pot life, since otherwise adhesive losses will occur due to an early adhesive curing. Therefore, a two-component adhesive should be mixed only after the adherends to be bonded have been matched according to the stipulated regulations and the surfaces have been prepared respectively.

3.2
Two-Component and One-Component Reactive Adhesives

The basic principle of curing of a reactive adhesive to form an adhesive layer is a chemical reaction, as described in Section 3.1. The given resin component A requires a "partner" for reaction. There are two different possibilities of "partner selection":

- Adhesives that have to be mixed with a second component for curing (two-component reactive adhesives).
- Adhesives that cure without the admixture of a second component, since the latter is already determined by the chemical state of the adherend surface or diffuses into the glueline, for example, water molecules (one-component reactive adhesives).

3.2.1
Two-Component Reactive Adhesives

In the case of *two-component reactive adhesives,* a second hardener component B is added to the resin component A in the mixing ratio prescribed by the adhesive manufacturer. Both components are then mixed evenly according to the processes described in Section 7.2.2 and applied to the adherends. The following adhesive types are characteristic for such two-component adhesives:

- cold-curing two-component epoxy resin adhesives (Section 4.1.1),
- cold-curing two-component polyurethane adhesives (Section 4.2.1),
- cold-curing two-component methacrylate adhesives (Section 4.3.3),
- other cold-curing two-component adhesives on polyester basis as well as special rubber types.

Cold-curing reactive adhesives are adhesives that enable strong and functional bonded joints due to the monomers' reactivity already at room temperature. Additional heat supply may contribute to a reduction of the curing time, especially in the case of epoxy resin adhesives, but need not necessarily to be done.

3.2.2
One-Component Reactive Adhesives

One-component reactive adhesives are applied to the adherends only in the form of one (the resin) component. That the curing of an adhesive layer still occurs is due to the fact that the reactive conditions required for the curing of the resin component in the glueline are given. Such conditions may be, for example:

- The water molecules present on the adherend surfaces that cause the polymerization (Section 4.3) of the *cyanoacrylate adhesives* (Section 4.3.1).
- The contact with metal surfaces that, in the case of *anaerobic adhesives* (Section 4.3.4), enables the reaction to an adhesive layer if, at the same time, the liquid adhesive (isolated by the adherends) is no longer in contact with the oxygen of the air.
- The water adsorbed on the adherends and present in the ambient air, which is available as a reactant for the moisture-curing *one-component polyurethane adhesives* (Section 4.2.2). This reaction is also typical for *sealing compounds* on polyurethane basis (Section 4.9) used, for example, in construction engineering for the sealing of joints between window and door frames and brickwork.
- In the case of *one-component silicon adhesives and sealing compounds* also humidity, which is responsible for the reaction to glueline, respectively, joint sealing. This reaction causes special silicones to develop acetic acid as a byproduct, which is perceptible from its typical smell.
- Another possibility of applying reactive adhesives in the form of only one component without mixing will be given, if the components A and B do not react with each other after mixing, because due to their chemical composition they are too

"low-reactive" at room temperature. Therefore, these adhesives can be stored in a mixed state at room temperature (or in a freezer chest to prolong storing time). After the application to the adherends, heat supply is required to overcome the "low reactivity" (Section 3.1.4). Such adhesives are called "blocked" reactive adhesives. The addition of "catalysts", effective only at higher temperatures, can also block such systems for a reaction at room temperature. Typical examples for this kind of adhesive are the *hot-curing epoxy resin adhesives* manufacturers already offer in mixed form in cartouches or as films (for extensive bonded joints, e.g., in aircraft construction).

One-component reactive adhesives have to be distinguished from the *physically setting adhesives* described in Chapter 5, which generally occur only in the form of one component, namely, the already "finished" polymer, for example, in the case of hot-melt adhesives, dispersion adhesives and solvent-based adhesives. They are called *one-component adhesives*.

3.3
Properties of Adhesive Layers

The applied liquid adhesive develops into the adhesive layer according to the chemical reactions or physical processes described in Sections 2.2.1 and 2.2.2 as well as in Chapter 3. So for reasons of preciseness, we talk of

- *adhesive,* as long as it is not cured; and
- *adhesive layer* after the curing of the adhesive.

Thus, in a finished bonded joint there is the adhesive layer and not the adhesive (see also Figure 1.3). Depending on the course of the chemical reactions the monomers can combine to straight and branched or even crosslinked polymer molecules (also called *macromolecules,* from Greek *makros* = large). These different polymer structures show considerably different properties, particularly under heat exposure.

3.3.1
Thermoplastics

Polymers form straight and partly even branched chains. At first, they soften (plastically) under heat supply and liquefy at rising temperature. After cooling down they solidify again. This property led to the name *thermoplastics* (also of Greek origin *thermos* = warm), that is, substances that soften or plasticize under heat. Typical examples are the hot-melt adhesives described in Section 5.1.

3.3.2
Thermoset Plastics

Polymers with crosslinked structures. They cannot melt under heat supply, since their individual chain segments are strongly chemically bonded (as, e.g., a wire mesh welded at the crossings). In Figure 3.5 such crosslink points are indicated by black dots. They are called thermoset materials. In contrast to most thermoplastics, they are also insoluble in organic solvents.

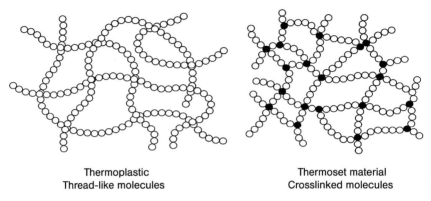

Figure 3.5 Thermoplastic and thermoset polymer structures.

The different behavior of thermoplastics and thermoset materials at increasing temperature is schematically shown in Figure 3.6. Depending on their chemical composition, softening and melting temperature of thermoplastic materials can vary to a great extent. Hot-melt adhesives, for example, are applied as melt within a temperature range of approx. 120–240 °C. In the case of thermoset materials the depicted temperature dependence depends much on the crosslinked state.

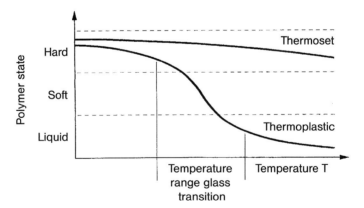

Figure 3.6 Temperature dependence of the polymer state of thermoplastic and thermosets (diagrammatic).

With continuously rising temperatures, a chemical decomposition of the polymers sets in, that is, the thermal decomposition of the molecule structures. This process is irreversible, which means it cannot be reversed by cooling down.

3.3.3
Elastomers

Polymers that, in contrast to thermosets, have a macromolecule structure with wide-meshed crosslinks are called *elastomers*. Their characteristic property is their not being flowable up to the temperature range of chemical decomposition, but they are rubbery-elastic and reversibly deformable, to a large extent independent of temperature (e.g., rubber products).

In addition to the criteria of the adhesive classification given in Section 2.2, another subdivision by the kind of adhesive layer or polymer properties formed by the adhesives is possible based on the aforementioned explanations (Figure 3.7, supplemented by typical adhesive raw materials).

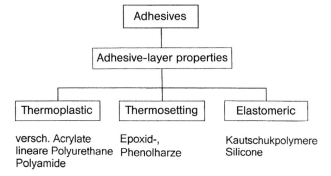

Figure 3.7 Subdivision of adhesives according to their kind of polymer properties.

3.3.4
Glass Transition Temperature

In connection with the behavior of polymers under the influence of heat the term *glass transition temperature* T_g is important, especially for thermoplastic material (Figure 3.6). This means a temperature range in which mechanical properties (e.g., strength, deformability) of a polymer change considerably.

Below the glass transition temperature (i.e. towards lower temperatures) the mentioned mechanical properties have higher values, which decrease partly very strongly when the glass transition temperature is exceeded (i.e. towards higher temperatures). The reason is the increasing mobility of the molecules given at higher temperatures, as already described in Section 3.1.4. Each individual polymer has its characteristic glass transition temperature value depending mainly on the molecule structure (linear, branched or crosslinked). Thus, values of, for example, rubber types are in the range of −50 to −70 °C, while epoxy resins can reach

100–120 °C. It is important to know the glass transition temperature of polymer adhesive layers when high-temperature strain is to be expected.

3.3.5
Creep

The peeling off of a hook fitted with a pressure sensitive adhesive and attached to a ceramic or glass surface can be regarded as a typical example for the creep behavior of an adhesive layer. In particular, thermoplastic adhesives that, to a great extent, also include pressure-sensitive adhesives (Section 5.6) tend to creep under high strain. A reason for this behavior is the time-related failure of individual bonds between the polymer molecules due to the strain imposed from outside. The application of adhesives with a higher crosslink ratio can reduce the adhesive layers' tendency to creep.

Complementary literature to Chapter 3:
[K4, M4, W3].

4
Important Reactive Adhesives

As already described in Section 3.2 there are two different kinds of reactive adhesives, that is,

- two-component adhesives,
- one-component adhesives.

In the following description of the most important reactive adhesives it will be particularly referred to their allocation to one of these two groups. Furthermore, additional advice will be given regarding the specific properties of the adhesives, their application conditions and essential chemical formulas.

4.1
Epoxy Resin Adhesives

4.1.1
Two Component Epoxy Resin Adhesives

The epoxy resin adhesives undoubtedly represent the most important group of reactive adhesives. The reason for this is the varied formulation possibilities such organic compounds provide to be able to offer "tailor-made" adhesives for the most varied fields of application. Characteristic of the structure of the monomers is a specific arrangement of carbon and oxygen, which, as part of the molecule, allows for high reactivity with other monomers. In this *epoxy group*, two carbon atoms and one oxygen atom have combined into a "triangle":

$$\boxed{A} \!-\! \overset{\overset{H}{|}}{C} \!-\! \overset{\overset{H\;H}{\diagdown\!\diagup}}{\underset{\diagdown\!\diagup}{\underset{O}{C}}}$$

Resin component A with epoxy group

The curing reaction is initiated by the admixture of compounds to the resin component as the second component B, which are called "hardeners" and that are able to open the "epoxy triangle":

Applied Adhesive Bonding: A Practical Guide for Flawless Results. Gerd Habenicht
Copyright © 2009 WILEY-VCH Verlag GmbH & Co. KGaA, Weinheim
ISBN: 978-3-527-32014-1

4 Important Reactive Adhesives

```
─[ A ]─C-C─
       | \/
       H  O-
       (with H H on carbons)
```

Thus, "bond arms" (valencies) develop to which the molecules of the hardener components can attach. Typical representatives of this group are the so-called amines:

```
─[ B ]─N⟨H
         H
```

Hardener component B with amine group

These two different monomers A and B with their respective "reactive" epoxy and amine groups can, according to the schematic in Figure 4.1, combine by a chemical reaction, they "add" according to the scheme A + B + A + B + … to the macromolecules that develop the hard (cured) adhesive layer. From this kind of polymer development the term *polyaddition* typical for this kind of reaction is derived.

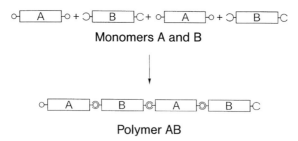

Figure 4.1 Schematic of a polyaddition reaction of two different monomers A and B to one polymer AB.

Since several reactive groups may exist at the respective monomer molecules A and B, net-like linked polymer molecules with thermoset adhesive layer properties occur in polyaddition. Due to the various selection possibilities regarding the chemical structure of the components A and B, curing behavior can be influenced as well. Differences are made between:

- curing at room temperature (cold curing),
- curing at elevated temperature (heat curing up to approximately 120 °C, hot curing up to approximately 250 °C), and
- application with short or long pot lives (minutes, hours, days).

▶ **Supplementary Information:**

- For the application of the two-component epoxy polyaddition adhesives described here, it is important to observe the mixing ratio indicated on the packaging (Section 3.1.2).

- One of the two components is often pigmented to ensure a uniform mixture. The mixing procedure has to be carried out until a uniform color of the mixture is obtained.
- Adhesives in double cartridges offer a particularly convenient means of application (Section 7.2.2.4).
- Packaging in the form of film wraps, where both components are separated by a sealed seam and the wraps already contain resin and hardener in the required mixing ratio, is very favorable for occasional application.
- When tubes are used, care has to be taken that the caps are not mixed up since otherwise they will be bonded to the tube.

4.1.2
One-Component Epoxy Resin Adhesives

One-component epoxy resin adhesives are almost limited to industrial use (car manufacture, aircraft construction, electronics). Resin and hardener components are already mixed, due to special formulations, however, they are prevented from reacting with each other at room temperature and thus from curing (blocking, Section 3.1.4).

So, one-component systems, for example for aircraft construction, are preferably produced as adhesive films, which have to be stored at low temperatures (up to approximately −20 °C). When they are cut to the dimensions of the adherends (compounding) and the adherends are fixed, they will be cured at high temperatures (approximately 140–160 °C).

4.1.3
Reactive Epoxy Resin Hot-Melt Adhesives

These at room temperature plastic/hard one-component systems are usually processed in adhesive melting drum installations. Adhesive heating occurs by means of heated panels that, depending on the consumption, continuously sink in the drum and heat the adhesive "in layers" until the curing reaction and viscosity reduction set in. The adhesive melt is pumped to the application appliances by thermally insulated hoses that are also heated. While the melt is solidifying, the final curing occurs in the glueline and results in an adhesive layer (just like a hot-melt adhesive, Section 5.1) due to the chemical reaction described in Section 4.1.1.

The advantage of such adhesive layers is their high long-term heat resistance compared to "normal" thermoplastic hot-melt adhesives, owing to their thermoset degree of crosslinkage.

The epoxy resin hot-melt adhesives are mainly applied in car manufacture, for example, in relining by bonding (engine bonnet, tailgate).

4.1.4
Properties and Application of Epoxy Resin Adhesives

The following properties of epoxy resin adhesives are regarded as fundamental:
- high strength, even at thermal stress due to their high degree of crosslinkage,
- excellent adhesion to almost all materials (for exceptions see Section 9.2 "Plastics"),
- high resistance to humidity,
- good ageing resistance towards environmental effects.

However, these positive properties are compensated by relatively limited deformation possibilities of the adhesive layers. This limits the bonding of flexible materials under continuous rolling or bending stress. For applications with special demands on deformation properties of the adhesive layers (e.g., in the case of crash stress in car manufacture) systems with special flexibilizating additives are available.

The processing of epoxy resin adhesives can be summarized as:
- two-component systems: mixtures of components (observe pot life!) – application to adherends – curing to final strength;
- one-component systems: application to the adherends – curing to final strength by heat supply;
- curing at increased temperature results in an increase in strength and stability of the bonding due to the higher crosslinkage.

Epoxy resins continue to be essential preliminary products for the manufacturing of fiber-reinforced compound materials with glass, carbon or plastic fibers imbedded. As far as their mechanical properties are concerned, they can compete with metal materials, for example, in car manufacturing, boat building and aircraft construction.

4.2
Polyurethane (PUR) Adhesives

The PUR (or PU) adhesives also cure according to the described mechanism of polyaddition. The reactive group at the molecules of resin component A has the following chemical structure and is called an "isocyanate group":

$$\boxed{\quad A \quad}\!-\!N\!=\!C\!=\!O$$

It is a property of this isocyanate group to react with compounds in which the reactive group, the so-called *hydroxide group* exists. Such compounds that are necessary for the isocyanate crosslinkage are called *polyoles*, since there are mostly several –O–H groups in the molecules (in organic chemistry, the –O–H group of certain molecule structures is indicated by the syllable ending "ol"). An example is the chemical compound "*alcohol*" for which this group is also characteristic. Therefore, in two-component PUR adhesives, polyoles are the hardener component B:

$$\boxed{-\boxed{B}-\text{O}-\text{H}}$$

The molecule arrangement arising due to the chemical reaction of A and B is called the *urethane group*. If several of these groups are contained in a polymer molecule, macromolecules, called *polyurethanes*, will develop, which, after the curing reaction, will finally represent the adhesive layer.

Owing to the various original compounds to which the reactive isocyanate and hydroxide groups are chemically bound, a variety of polyurethane adhesives are available, which will be briefly described below. A classification is to be found in Figure 4.3.

4.2.1
Two-Component Polyurethane Adhesives (Solvent-Free)

In these adhesive systems

- component A consists of a low-molecular weight polyisocyanate,
- component B consists of a low-molecular weight polyol.

As these components consist of only relatively small molecules, the viscosity is low so that they can be easily mixed by stirring according to the prescribed mixing ratio. The curing reaction usually occurs at room temperature.

Supplementary information regarding these systems is to be found at the end of Section 4.2.2.

4.2.2
One-Component Polyurethane Adhesives (Solvent-Free)

The main component of these adhesives consists of pre-crosslinked, high molecular weight polyurethanes. These prepolymers (Section 2.1.3) exist in a liquid or paste-like state and still show free isocyanate groups (so-called polyisocyanate-polyurethane). With these isocyanate groups the –O–H group

$$\text{H}-\text{O}-\text{H}$$

contained in the water is able to react. Thus, water molecules are important as the second component for the final crosslinking. They exist as reactants:

- in the form of relative humidity. This is defined as the ratio of the amount of water vapor contained in the air and the maximum amount possible at the respective temperature. At a temperature of 20 °C and a relative humidity of 70% one cubic meter air contains 12 g of water;
- adsorbed at the adherend surfaces; or

- absorbed in the adherends (e.g., in the case of wood, leather, cardboard, paper, plastic foams, masonry).

In special cases the water required for crosslinkage can be applied by spraying onto the adhesive layer, not yet cured, before fixing the adherends. However, this procedure requires certain care, since in the case of excessive dosage carbon dioxide develops as a byproduct that, if gas-impermeable adherends are concerned, leads to the bulging of the bonded joint and requires regular pressure application. This is a common process in bonding foam composite elements with external metal sheets and/or wood/synthetic resin laminate (e.g., in caravan and container building). Figure 4.2 shows the possible sources of moisture.

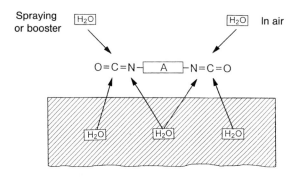

Component A: PUR prepolymer with reactive isocyanate groups

Component B: Water

Figure 4.2 Moisture curing of one-component polyurethane adhesives.

Since these one-component polyurethane adhesives cure through reaction with water, one talks of *moisture curing one-component polyurethane adhesives*. Their characteristic feature is their excellent adhesiveness. Apart from adhesive applications, these formulations are applied to a great extent as sealing compounds, for example, as fitting foam glue or PU-foam (Section 4.9).

▶ **Supplementary Information:**

- Such adhesives are not suitable for extensive bonded joints of moisture impermeable materials with regularly applied adhesive layers, since curing is not possible. In such cases the two-component systems or, if possible, additional moisturization, as mentioned before, is recommended.
- Since curing will be possible only if sufficient humidity from the air or from humid adherends gets into contact with the adhesive, the adhesive should be applied in dots, in the form of parallel lines or waves according to the size of the adhesive layer. A spiral line application requires very long curing times, since inside, the adhesive applied in spirals form cannot absorb sufficient moisture.

- For moisture curing one-component PUR adhesives, the industry even offers booster systems working with moisture-containing gels. This enables an accelerated curing independent of the prevailing humidity level.
- Adhesive foams applied out of spray cans are available on the same basis as the one-component polyurethane adhesive. Due to the contact with humidity in the air or moisture of the adherends, an inherently stable, well-adhering foam develops that shows excellent sealing properties.
- The application of these adhesives is advantageous as an alternative to mechanical joining procedures such as plugging, screwing, nailing, and so on, as well as to many applications around the home (signs, mirrors, tiles, panel walls, letter boxes).
- **Tip:** After bonding/sealing put a match into the still liquid mass at the outlet of the can valve to be able to pull out the cured foam plug before the next application.

4.2.3
Reactive Polyurethane Hot-Melt Adhesives (Solvent-Free)

This kind of adhesive, frequently applied in the car industry, for example, for bonding in place of window panes (direct glazing), consists of high-molecular weight, fusible polyisocyanates with terminal reactive isocyanate groups in a state of very high viscosity at room temperature. Thus, prior to application, melting to approximately 60–80 °C is required to enable the application to the panes by means of compressed air out of nozzles. After cooling down in the glueline between the body flange and window pane, the adhesive layers show such a high cohesive strength that the vehicle can be moved in-house. Since the adhesive layer still contains free isocyanate groups, these groups are able to react with the humidity in the air to achieve the final crosslinkage. Depending on the humidity and based on the water molecules' small contact surface on the edges of the adhesive layers, this procedure is rather time dependent; it may last for days. This additional crosslinkage results in adhesive layers showing considerably higher thermal resistance compared to "normal" hot-melt adhesives (Section 5.1). This procedure with different curing mechanisms (cooling down – crosslinking) led to the name *reactive hot-melt adhesives*.

During the application, the so-called skinning time has to be observed, that is, the time in which the contact with the humidity of the air already leads to a curing reaction on the surface. This "skin" prevents a sufficient wettability of the second adherend.

The outstanding property of such PUR adhesive layers is their high elasticity or flexibility over a wide temperature range (approximately –40 to 80 °C). This thermomechanic behavior is the precondition for the application in car industry.

4.2.4
One-Component Polyurethane Solvent-Based Adhesives

Here, physically setting adhesives are concerned. The already crosslinked hydroxyl-polyurethane macromolecules are dissolved in organic solvents. Depending on the structure of the materials to be bonded, the solvents must evaporate completely or for the most part prior to the fixing of the adherends. The remaining –O–H groups on the macromolecules are one decisive precondition for the very good adhesive strengths on the adherend surfaces.

4.2.5
Two-Component Polyurethane Solvent-Based Adhesives

As far as component A is concerned, these adhesives consist of a hydroxyl-polyurethane dissolved in a solvent, with a polyisocyanate, also in the liquid state, admixed as component B. The crosslinkage thus possible enables higher cohesive strength of the adhesive layer compared to the abovementioned one-component systems, and therefore even higher resistance towards chemical and physical stress.

4.2.6
Polyurethane Dispersion Adhesives

In contrast to solvent-based systems, the PUR dispersions (Section 5.4) are characterized by their incombustibility and thus by considerably smoother processing. They are high molecular weight hydroxyl polyurethanes disperged in water. Apart from the physically setting one-component systems, two-component systems are also applied, with component B containing special polyisocyanates that react with the –O–H-groups of the hydroxyl polyurethane in aqueous solution.

Figure 4.3 summarizes again the overview of the described polyurethane adhesives.

The described polyurethane adhesives are applied depending on the materials to be bonded and the given application conditions in the different fields of industry, as for example

- in the shoe industry for the bonding of soles to the upper parts;
- in the packaging industry for the lamination (extensive bonding) of films of polyethylene, polyester, cellulose film, aluminum, paper, cardboard;
- in the car industry for bonding of composite constructions of polyurethane or polystyrene foam with top coats of wood, plastic, aluminum, sheet steel for vehicle bodies and for the bonding in place of window panes in the bodywork;
- for the bonding of flexible materials if permanently exposed to bending and rolling stresses (e.g., conveyor belts).

Special advantages of polyurethane adhesives are their excellent adhesion strength on many surfaces even on those otherwise difficult to bond, for example, flexible

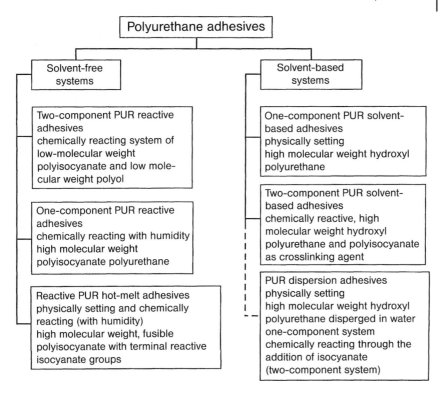

Figure 4.3 Classification of the polyurethane adhesives.

PVC. Furthermore, they are characterized by good chemical and thermal resistance as well as by high flexibility even at low temperatures. According to the polymer molecules' degree of crosslinkage, which depends on the respective chemical raw material, and thus determines, *inter alia*, the strength of the adhesive layer, PUR adhesives can cure to elastomers or thermoset materials.

4.3
Acrylic Adhesives

Regarding their curing reaction these adhesive systems differ considerably from the described epoxy and polyurethane adhesives that are characterized by the principle of polyaddition.

The special feature of the acrylates is the existence of the carbon–carbon double bond already mentioned in Section 2.1, that is, two carbon atoms are bound with each other by two "bonding arms" *(valencies)*:

$$-\boxed{\text{A}}-\overset{|}{\text{C}}=\overset{|}{\text{C}}-$$

In this *carbon–carbon double bond* one of the two bonds can be separated, thus creating two new bonding possibilities:

$$\text{—[A]—}\overset{|}{\underset{|}{C}}\text{—}\overset{|}{\underset{|}{C}}\text{—}$$

This "decomposition" of the double bond takes place in the curing reaction of the acrylates on a multitude of monomer molecules under certain conditions that will be described later. The monomers can combine to a polymer on each newly formed bond:

$$\text{—[A]—C—C—} + \text{—[A]—C—C—} + \text{—[A]—C—C—}$$

$$\downarrow$$

$$\text{—[A]—C—C—[A]—C—C—[A]—C—C—}$$

Consequently, this kind of reaction is not based on two monomer molecules A and B of completely different structure, as in the case of epoxy resin and polyurethane adhesives, but on monomers of the same kind or, at least as far as the C=C double bonds are concerned, of similar monomers. The double bond is therefore the precondition for the curing process of acrylate adhesives.

For the reaction mechanism described above the term *polymerization* has become established. Adhesives curing in this way are therefore called *polymerization adhesives*.

Thus, following Figure 4.1, this kind of reaction can be schematically depicted as follows (Figure 4.4):

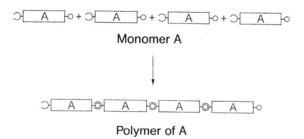

Monomer A

↓

Polymer of A

Figure 4.4 Schematic depiction of a polymerization reaction of monomers A of the same kind or of similar monomers A to a polymer of A.

Different acrylate adhesives are differentiated depending on the conditions the decomposition of the C=C double bond is subjected to. The cyanoacrylate adhesives are the best known of this group.

4.3.1
Cyanoacrylate Adhesives

These adhesives became generally known as *superglues*, since they cure within a very short time (within a range of seconds). Even if the syllable "cyano" leads to the assumption that something "toxic" is involved, this is by no means the case owing to the chemical structure of these substances. However, as applies to all adhesives, certain precautionary measures in application should be taken, which will be separately discussed in Section 7.5.

The trigger for the decomposition of the C=C double bond of cyanoacrylates is the moisture in the air that precipitates on the surfaces of the adherends, that is, which is "adsorbed" on them. As soon as the liquid adhesives get into contact with water molecules, the curing reaction in the glueline in the form of polymerization sets in at very high speed so that a "handling strength", that is, a bond strength is given that allows the further processing in the work flow. However, depending on the given moisture, the final strength develops only after some hours.

The consequence of the quick reaction with moisture is the adhesive's need to be stored in absolutely leakproof containers, mostly plastic bottles, to prevent a curing reaction in the bottle.

At this point, an important difference compared to moisture curing, one-component polyurethane adhesives described in Section 4.2.2 has to be mentioned. While for cyanoacrylates already small traces of moisture are sufficient for a quick polymerization, polyurethanes require a considerably higher moisture amount for complete curing, because in such bonds, water in chemically bound form becomes a component of the adhesive layer. In the case of cyanoacrylates, water is merely the "starter" of the curing reaction.

Cyanoacrylate adhesives are applied in various fields of industry, for example,

- for the bonding of plastics (in the case of polyethylene and polypropylene with the respective primer, Section 9.2.6), rubber and rubber compounds;
- as tissue adhesive and spray bandages in medicine;
- for the bonding of electronic and optical components;
- for fixation joints.

▶ **Supplementary Information:**
- Since the effectiveness of the water adsorbed on the adherend surfaces only suffices for the polymerization of adhesive layers with limited thickness, such layers should not exceed 0.2 mm. Furthermore, the relative humidity of the processing rooms should range between approximately 40 and 70%.
- After the application of the adhesive, the adherends have to be fixed immediately, since otherwise the beginning polymerization will lead to a reduction of the bonding strength.
- Such adhesives are particularly suitable for smaller adherend surfaces, since for large-surface applications the open assembly time is not sufficient, thus curing may occur before the adherends are fixed.

- In dependence on the application, superglues are available in different processing forms – from easy-flow for an easy penetration of the adhesive in the case of small crevices, up to gel-like, pasty formulations for bondings on vertical surfaces and adherends with porous surfaces.
- When using cyanoacrylate adhesives it should be considered that the resistance of the bonded joints against water, especially at higher temperatures, is not comparable to that of epoxy joints. In the case of common household joints, unlimited "dishwasher safety" cannot be taken for granted. Heat resistance is usually limited to approximately 80 °C.
- Direct contact of adhesive and skin area (e.g., finger tips) must be strictly avoided, since skin moisture provokes immediate agglutination. Immerse agglutinated skin instantly into warm soapy water and try to detach the hands by moving them slowly. Afterwards grease your hands with skin cream. If adhesive splashes get into the eye, lacrimal fluid will cause immediate curing. Rinse the eye at once and consult an ophthalmologist. Working with protective glasses prevents such accidents!
- Keep superglues out of children's reach. The rapid curing of superglues represents a considerable risk potential.
- **Tip:** At low humidities, simply breathe on the adherend surface. The thus increased humidity on the surface accelerates the polymerization process.
- **Tip:** Leakproof sealing of an opened bottle is achieved by immersing the bottle top into liquid wax or stearin of a candle (after having extinguished the flame!). Once the stearin has cooled down, a leakproof cap develops that can be easily removed or penetrated for further use. Moreover, it is recommendable to keep the bottle in the fridge in a tightly closed glass to limit the moisture pick up to a minimum.

4.3.2
Radiation-Curing Adhesives

Another possibility of decomposing a C=C-double bond is supplying energy. Radiation contains energy in different forms; the heat radiation of a heater or an electric light bulb is known as well as the radiation of an X-ray tube able to penetrate body tissue. Particularly suitable for curing, thus the polymerization of the C=C-monomers, is *ultraviolet radiation* (UV-radiation), known as a part of solar radiation. If these ultraviolet rays meet with the adhesive monomers to which so-called photoinitiators are added, the double bonds included in the monomers will be decomposed and polymerization proceeds in a way similar to that described in Section 4.3.

Therefore, bonding with radiation-curing adhesives requires at least one of the adherends to be permeable to ultraviolet radiation. For this reason such adhesives are used for glass–glass and glass–metal joints (Section 9.3) or for UV-permeable plastics, for example, Plexiglas joints. A further wide range of applications is the manufacturing of adhesive tapes. The adhesive layer applied to the substrate is directly accessible to UV-radiation.

Apart from a suitable UV-curing adhesive, suitable UV-radiation sources are required, too, which are available in the form of flashlights or continuous lines. To achieve optimum curing, it is very important to adapt the UV-radiation precisely to the adhesive to be cured (Section 9.3.3).

Due to their photosensitivity, UV-curing adhesives should be kept in a dark place although they are sold in light-tight packings. The advantages of radiation-curing adhesives are their very short curing time (short clock cycles in manufacturing) and their one-component application.

4.3.3
Methacrylate Adhesives

These two-component reactive adhesives are characterized by a large scope of processing possibilities, since the pot lives are variable over a large time span. The resin basis is an acrylate, more precisely a methylmethacrylate, which is the same raw material as Plexiglas is made of (chemically: polymethylmethacrylate). In view of their chemical composition, the developing adhesive layers can be regarded as "Plexiglas layers", thus showing thermoplastic properties.

The acrylate monomers polymerize because of the C=C double bond contained in the molecule under the influence of a hardener characterized by an oxygen–oxygen bond, a so-called peroxide. (Even the hydrogenperoxide, a known bleaching agent, shows such a O–O bond). The triggering of the polymerization requires an accelerator already incorporated into the resin component by the manufacturer.

The methacrylate adhesives are processable according to the following methods:

1. The peroxide hardener is added to the resin in the form of powder. Very small quantities of approximately 1–3% are sufficient. The manufacturers offer "quick" adhesives with pot lives in the range of minutes and "slower" adhesives with pot lives up to one hour. Quick adhesives are usually applied by means of dosing systems. This is also recommendable in case of occasional bonding procedures and if mixing and dosing systems are not available, however, they have to be applied immediately after mixing (MIX-system).
2. This variation implies that the peroxide hardener dissolved in an organic solvent will be applied to one of the two adherends. After the evaporation of the solvent, the hardener remains on the surface in a very thin layer where it can remain for a sufficiently long time without changing. The resin component provided with the accelerator, which is not subjected to pot-life limitation, will be applied to the other adherend. Only when both adherends are fixed does the contact between hardener and resin/accelerator system lead to the chemical reaction of the adhesive layer formation. This procedure is common in industrial use, but also advantageous for trade and semi-industrial use, since there are no pot-life limitations. Care has to be taken that the adhesive layer is not too thick since otherwise the amount of hardener applied in a thin layer to one of the adherends is not sufficient for complete curing. Since the hardener is applied

in form of a solution, this procedure is called the *hardener-varnish procedure* (NO-MIX system).

3. This is the so-called *A-B procedure*. Its advantage is that the addition of a hardener, required only in low concentrations (approximately 3–5%) – which may lead to mixing problems, respectively, to high investment costs occurs already at the adhesive manufacturer's facility.

   Component A consists of the acrylate resin with the accelerator, component B consists of resin without the accelerator, however with the addition of the hardener. Both components are storable without pot-life limitation.

   Processing occurs by mixing the components A and B. The accelerator contained in component A together with the hardener in component B trigger off the reaction. The user can choose between AB adhesives with different pot lives. Another possibility is the application of component A to one adherend and of component B to the other. After the fixing of the adherends both components mix, which results in the curing of the adhesive layer. Alternatively, the components A and B can be applied one over the other in the form of two adhesive lines. Then, immediate joining follows and curing starts at once.

All three methods are carried out at room temperature. Figure 4.5 shows additionally how to proceed when working according to the A-B method.

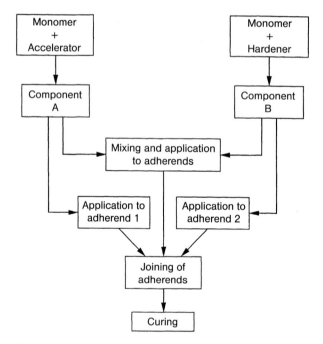

Figure 4.5 Processing of methacrylate adhesives according to the A-B method.

Consequently, the basis of methacrylate adhesive processing according to the A-B method is the following principle:

- no reaction between accelerator and monomer,
- no reaction between hardener and monomer,
- reaction occurs only when accelerator + hardener + monomer are combined in a mixture.

Methacrylate adhesives are perfectly suitable for bonded joints of metals, glasses and thermosetting plastics. They have proven their worth especially in loudspeaker manufacturing in the bonding of ferrite cores. Particularly advantageous are the high strengths as well as the short curing times without high expenditures for mixing units.

▶ **Supplementary Information:**

- In this context the essential difference of methacrylates compared to the processing of two-component epoxy resin and polyurethane adhesives has to be discussed again. While in the case of such adhesives, the second component, usually in liquid or pasty form, is mixed with the resin component in a ratio of approximately 1 : 1, the hardening powder in relation to the given methacrylat-resin component is dosed in a very small percentage when processed according to method 1 described in Section 4.3.3. This difference is justified by the special curing mechanism of the methacrylate adhesives.
- For method 3 (A-B procedure) double cartridges are available, too, which enable processing (Section 7.2.2.4) according to the suitable mixing ratio.
- Acrylates in certain formulations are suitable for construction purposes in construction engineering. The reaction products that develop after curing show high pressure resistance, they can be processed by boring, milling, grinding and sawing and the like, and due to their gap-filling properties they are used for joint repair (polymer mortar, Section 4.10).

4.3.4
Anaerobic Adhesives

These one-component reactive adhesives are liquid as long as they are in contact with the oxygen of the air, which prevents the monomer molecules from polymerizing. Only after the joining of the adherends, for example, after the application of a screw nut onto a screw thread coated with the adhesive, can oxygen no longer come into contact with the adhesive, thus it will cure. In this chemical procedure, the metal surfaces of the adherends play an important role as well; therefore the adhesive layers should not exceed a thickness of 0.2 mm. Deriving from the Greek word "anaerob" (living without oxygen), these adhesives are called *anaerobic adhesives*.

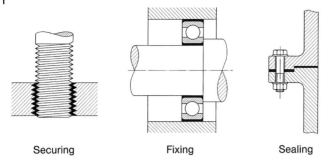

Figure 4.6 Application possibilities of anaerobic adhesives.

Anaerobic adhesives mainly serve the purpose of protecting threads from undesired loosening as a consequence of vibrations or dynamic stress and represent a preferred alternative to mechanical locking devices. Furthermore, they enable shaft–hub connections, for example, when mounting a gear wheel on a shaft (Figure 11.6). They have also delivered optimal performance as liquidly applied surface seals alternatively to the solid seals required in the respective dimensions. Figure 4.6 shows these application possibilities.

Depending on their chemical basis and the resulting thermosetting crosslinkage ratio, the adhesive layers are thermally resistant and thus perfectly suitable for transmission and engine manufacturing. If repair is required, the bonded joints can be removed by heating them to approximately 120–150 °C.

▶ **Supplementary Information:**

- The adhesive is applied to the fat-free thread before the screw is screwed in immediately. Sufficient initial strength sets in after approximately 30 minutes, the final functional strength after approximately 3 hours. Fixing by means of adhesives has the additional advantage of absolute tightness of the screw connection, and it prevents the possible rust formation within the boundary layer.
- In the case of blind holes, apply adhesive to the lower third of the borehole so that the adhesive will be pressed up on the (inner) walls of the threads when the screw or the stay bolt is tightened or even when a pin is bonded in place.
- For cases in which such secured screw connections have to be loosened again, adhesives with different adhesive-layer strengths (initial breakaway torques) are available.
- Of course, when bonding plastic the metal contact important for the curing of such adhesives is not given. For this reason, manufacturers have so-called activators in store that are applied to the plastic surfaces to enable curing.
- The relatively high portion of air and thus of oxygen in the headspace of the bottles is necessary to keep the adhesive liquid.

4.4
Unsaturated Polyester Resins (UP-Resins)

Such compounds – although as a functional group, they also have C=C-double bonds – are not assigned to the acrylate adhesives. These products are mentioned, because experience shows that repairs of boats, vehicle body work (e.g., caravans) and other plastic parts are frequently carried out, where unsaturated polyester resins as two-component systems with hardener component on styrene basis play an important role.

In this context, unsaturated resins are usually compounds containing C=C double bonds that, by polymerization and crosslinkage with the respective monomers, can be transferred into saturated thermosetting compounds that no longer contain double bonds. Here again the basis are resin and hardener components that cure after mixing at room temperature or in the heat.

In Section 7.3.2.1, the performance of repair work is described.

4.5
Phenolic Adhesives

Apart from polyaddition and polymerization adhesives, a third kind of adhesive that shows a special reaction mechanism during the curing process exists. They should be mentioned for the sake of completeness, although they are less important compared to the systems mentioned so far. Their characteristic is the fact that a byproduct arises in the formation of the polymers from the monomers, which has to be considered in curing. The central molecule of these adhesives is *formaldehyde*

$$\begin{array}{c} H \\ \diagdown \\ C=O \\ \diagup \\ H \end{array}$$

which reacts with other molecules, for example, phenol, urea, melamine under separation (condensation) of water molecules and develops an adhesive layer. From this process the name *polycondensation adhesives* derives (Figure 4.7).

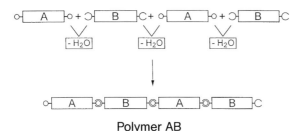

Polymer AB

Figure 4.7 Schematic diagram of a polycondensation reaction of two different monomers A and B to one polymer AB under separation of water.

Due to this water condensation, such adhesives have to be cured at high temperature and under high pressure in so-called autoclaves, when impermeable materials are bonded, to avoid an increase in volume of the adhesive layer caused by the water-vapour expansion.

Polycondensation adhesives based on phenol or phenol derivates with formaldehyde are mainly used in the production of wooden constructions (in woodworking one traditionally talks of "glueing" instead of "bonding"), for example, in the manufacturing of laminated bonds (plywood, chipboard, fiber board, beams). Since in these applications, the adherends are able to absorb the water developing during the polycondensation reaction, the processing occurs in heated presses to enable the application of the necessary contact pressure at the required temperature (Section 7.2.4). For applications in nonindustrial fields these adhesive are practically unimportant.

A special feature of phenolic adhesives is the extremely high heat resistance of the adhesive layers, up to several hundred degrees centigrade. Therefore, they are widely used in car manufacturing for the bonding of brake discs and clutch linings on metal carriers.

4.6
Silicones

The adhesives curing after polycondensation reaction also include silicones. Here, systems between organic and inorganic compounds (Section 2.2.5, Figure 2.4) are concerned. In their basic structure they have silicon–oxygen bonds instead of carbon chains:

$$-\overset{|}{\underset{|}{Si}}-O-\overset{|}{\underset{|}{Si}}-O-\overset{|}{\underset{|}{Si}}-O-$$

Their main field of application are sealing compounds (silicone rubber) for which they are sold in the form of reactive one-component systems (mainly in cartridges) (RTV-1-systems, room-temperature vulcanization). Just like the one-component polyurethanes described in Section 4.2.2, they cure under the influence of moisture from the ambient air. In certain formulations, this reaction leads to the separation of acetic acid perceptible by its characteristic odor. Adhesive and sealing layers on a silicone basis show the following characteristics:

- high thermal resistance up to 200 °C,
- very high flexibility even at low temperatures (–50 to –70 °C),
- excellent weathering resistance.

▶ **Supplementary Information:**

- Application as RTV-1 systems requires preconditions comparable to moisture-curing one-component polyurethane adhesives (Section 4.2.2). In order to ensure that curing is possible only with sufficient moisture, care has to be taken

that the adhesive and sealant in line form have enough air contact during the application (back ventilation, do not apply spiral or closed adhesive lines).
- If these preconditions are not given due to structural reasons, an alternative is the two-component system (RTV-2) with shorter curing times.
- The adherends have to be fixed immediately after the adhesive is applied, since otherwise early curing (skin formation on the adhesive surface) will set in causing a deterioration of the adhesive properties.
- Since the moisture penetrates the applied adhesive layeres by means of diffusion, the curing times range from hours to several days, depending on the joint geometry. In general, it can be assumed that curing from outside into the interior of the adhesive proceeds at approximately 2 mm per day.

4.7 Summary Reactive Adhesives

The reactive adhesives described in Sections 4.1 to 4.6 are summarized in Figure 4.8 according to their respective curing reaction.

With regard to the thermomechanical properties of their adhesive layers, the reactive adhesives mentioned before can be classified as follows:

- *Thermoset materials:* Epoxy resins, phenolic resins, polyurethanes (highly crosslinked), anaerobic adhesives.

- *Thermoplastics:* Cyanoacrylates, methacrylates, radiation-curing adhesives, polyurethanes (depending on the degree of crosslinkage).

- *Elastomers:* Silicones, polyurethanes (depending on the degree of crosslinkage).

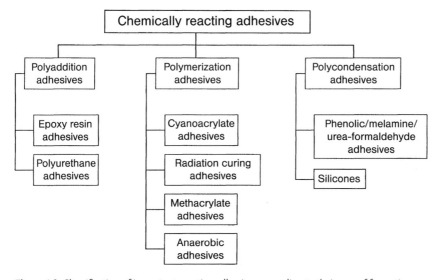

Figure 4.8 Classification of important reactive adhesives according to their way of formation.

The described reactive adhesives are suitable for bonding of nearly all metal and nonmetal materials used in industry, trade and even in the private sector. They are characterized by good to very good adhesive properties on correspondingly prepared surfaces (Section 7.1.2) as well as by stress-related strengths. Some plastics, especially polyethylene, however require special measures regarding their surface pretreatment. Here, we refer to Section 9.2.

Reactive adhesives in a broader sense include also

- film adhesives,
- sealing materials and polymer mortars,

which are described in the following sections, since their curing mechanisms are based on the chemical reactions already described.

4.8
Film Adhesives

Film adhesives have to be strictly delimitated from adhesive tapes and adhesive strips (Sections 5.6 and 5.7). Mainly blocked two-component reactive adhesives (Section 3.1.4) are used raw materials. For transport and storage (at low temperatures) they are applied to a – nonadhesive – substrate. Prior to processing they are removed and then applied between the adherends (compounding) and cured under pressure and heat (Sections 3.1.4, 3.2.2 and 4.1.2). Special film adhesives (e.g., phenolic resin nitrile rubber) are also activated by suitable solvents.

4.9
Sealing Materials

With regard to their chemical structure and their curing reactions, sealing materials are closely related to adhesives. The most frequently applied sealing materials are based on

- polyurethanes (Section 4.2),
- silicones (Section 4.6),
- MS-polymers (polymers with moisture-curing **m**odified **s**ilane groups),
- polysulfides,
- elastomers with different chemical bases (e.g., butylenes).

Important properties of these raw materials are the elasticity behavior given over a wide temperature range in cold and heat and the excellent aging and weathering resistance. Thus, various applications, above all outdoors, are possible. They are applied in nearly all fields of industry, for which special formulations are available according to the requirements and the material combinations to be sealed. The manufacturers show partly very detailed information on these products, which has to be referred to at this point.

With regard to processing, sealing materials are subdivided as follows:

- *Liquid sealings:* Sealing materials applied to the sealing joint in thin to viscous consistency. If assembly of the components provided with the sealing material takes place *prior* to the completed crosslinkage of the polymers, it is talked of *wet bonding* or of the *formed-in-place-gasket-(FIPG) method*. In the case of *dry bonding,* the assembly of the sealing parts occurs only *after* the completed crosslinkage of the sealing material. This method is called *cured-in-place-gasket-(CIPG) method*.

- *Foam sealings:* These sealings contain small air bubbles that allow the compression of the sealings. In this system the actual sealing function occurs only in the compressed state.

- *Compact sealings:* They are given in a nonporous state and are compressible only to a minor degree. O-rings are to be mentioned as examples, which are producible at the point of application according to the required diameter by means of cyanoacrylate adhesives.

- **Tip:** Use a new blade for cutting the O-ring to size, otherwise an uneven cut with nonparallel cut surfaces may arise. Do not use scissors for the same reason. Both ends of the O-ring have to be cut anew due to possible surface contamination during storage.

4.10
Polymer Mortars

These materials, in a broader sense not to be assigned to the adhesives, are mainly used in the construction sector for repairs, for the stabilization of anchors in boreholes as well as for reconstruction purposes. Here, mortar is concerned that contains liquid-reactive synthetic resins as a complete system or as additives instead of cement as conventional binding agent. Common additives in concrete technology are silica flours and silica sands. Compared to cement mortar, polymer mortar is characterized by high chemical resistance, higher tensile strength, lower modulus of elasticity as well as a shorter setting time. Epoxides, polyurethanes, unsaturated polyesters and methacrylates are used as synthetic resins.

Mortar masses on a purely inorganic basis consist of hydraulically setting components such as cement or gypsum that harden with water. In contrast to these products, there is nonhydraulically, that is, air-hardening mortar.

Complementary literature to Chapter 4:
[B5, G1, H1, M1, W1].

5
Physically Setting Adhesives

As already described in Section 2.2.2, these adhesives do not provoke chemical reactions in the glueline, since the adhesive layer polymers are already in a "ready state". Thus no second component is added to the adhesives prior to their processing – they are, without exception, one-component systems. To enable their application to the adherends, they have to be transformed to a wettable state (Section 6.2). Such transformation possibilities are echoed in the names of some adhesives.

5.1
Hot-Melt Adhesives

Hot-melt adhesives, which belong to thermoplastics (Section 3.3.1), are liquefied by heat supply, for example, in electrically heated nozzles of the application device, and then applied to the adherends. Since the hot-melt cools down very quickly the adherends have to be affixed immediately. The "open assembly time" of these adhesives, that is, the period of time between the application of the adhesive and the fixing of the adherends, is very short and must not be exceeded. The open assembly time strongly depends on the heat conductivity properties of the adherends; the faster they dissipate the heat from the melt, the shorter is the open assembly time. Hot-melt adhesives are available in the form of blocks, rods, films, granulate or even in the form of powder.

According to the chemical structure of the hot-melt adhesive polymers (polyamide resins, saturated polyester, ethylene vinyl acetate copolymers, polyurethanes), the processing temperatures range between 120 and 240 °C.

During the processing phase of heating–cooling, the hot-melt adhesives do not undergo chemical changes. In contrast to chemically reactive and solvent-containing adhesives, they show some remarkable advantages:

- solvent-free and thus no special fire protection measures required;
- processable as one-component systems;
- very short setting times, therefore high production rates possible;
- as thermoplastics they offer the possibility to separate (disconnect) bonded joints by heat supply (important in view of recycling).

Applied Adhesive Bonding: A Practical Guide for Flawless Results. Gerd Habenicht
Copyright © 2009 WILEY-VCH Verlag GmbH & Co. KGaA, Weinheim
ISBN: 978-3-527-32014-1

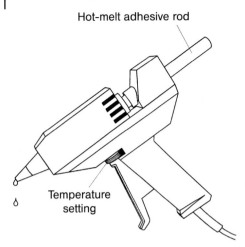

Figure 5.1 Manual hot-melt adhesive applicator.

Processing is carried out by means of electrically heated manual pistols with low consumption in which the hot-melt adhesive, in the form of a rod, is inserted. The activation of a feeding device causes the rod to be pressed into the heating zone in which it melts and discharges from a nozzle (Figure 5.1). For serial bonding, melting systems and related automatically working dosing units are available.

Due to their relatively simple processing, hot-melt adhesives are widely applied, for example in the

- packaging industry (cardboard bonding),
- book and folder production (spine gluing),
- wood and furniture industry (edge gluing, veneer coating, structural bondings),
- shoe industry (sole and internal bonding),
- electronic engineering (wire-wound coils, fixation of wires),
- textile industry.

In the latter case, the hot-melt adhesives are applied in the form of films (also perforated for moisture exchange) as *heat-sealing adhesives*. The film is put between the two webs to be bonded and is then melted in heating presses or by means of an iron. The melt penetrates the tissue and, after cooling, forms a firm bond, thus stiffening the tissue.

▶ **Supplementary Information:**

- Due to the considerably higher heat conductivity of metals in contrast to wood or plastics, it is advisable to preheat the adherends to be bonded to the temperature of the melt in order to achieve good adhesive strengths. (Hot-air gun, where appropriate, hair-dryer at highest heat level; because of its electric conductivity never use a microwave oven!)

- Owing to their limited thermal resistance, bonding of plastics, in particular of thermoplastics, requires certain precautionary measures to avoid deformation of the adherends (use hot-melt adhesives with low processing temperature, e.g., on a polyamide basis).
- Due to the limited open assembly time, large area bondings are only possible on a restricted basis. Special hot-melt adhesives are available for the bonding of larger surfaces for industrial purposes, and especially adjusted applicators are utilized (spray application).
- Beware of burns – in liquid state, melts have temperatures of the order of 200 °C.

5.2
Solvent-Based Adhesives

Solvent-based adhesives are adhesives with polymers dissolved or pasted in organic solvents. The solvents or solvent mixtures are only processing aids and have to be removed, either partly or completely, from the applied liquid adhesive layer through evaporation or penetration prior to the fixing of the adherends. The first case is necessary for solvent-impermeable materials (metals, glass, thermosetting plastics), the second case concerns porous and solvent-permeable materials (paper, cardboard, wood, leather). This process can be accelerated by heat supply. Solvents are mainly esters, ketones, if applicable, portions of different alcohols. The total solvent portion ranges between 75–85%.

The following polymers or polymer mixtures, respectively, in combination with tackifying resins are mainly used for solvent-based adhesives:

- polyvinyl acetate and copolymers,
- polyvinyl alcohol and copolymers,
- natural and synthetic rubbers,
- nitrocellulose,
- acrylates,
- polyurethanes.

Important terms for the processing of solvent-based kinds of adhesives are explained below (Figure 5.2):

- *Minimum drying time:* The largest part of the solvents contained in the liquid adhesive after application evaporates during the minimum drying time. This time should pass in any case prior to the fixing of the adherends in order to achieve a high initial strength as soon as possible.

- *Maximum drying time:* This is the period of time which only just enables a bonded joint. If the maximum drying time is exceeded, the polymer layers will already have solidified to such an extent that the cohesion strength of the adhesive layer may be affected. An exception are the contact adhesives (Section 5.3)

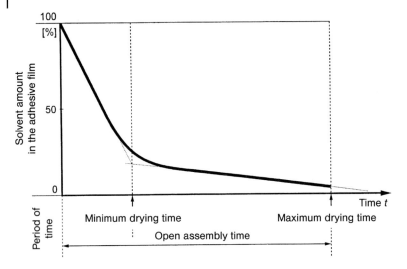

Figure 5.2 Time dependence of the solvent amount in the liquid adhesive layer.

that still show tackifying components in their formulation and where high surface pressure is applied after the fixing of the adherends.

- *Open assembly time:* Period of time, also called "wet bonding time", which may pass between the application of the adhesive and the fixing of the adherends. Deterioration of the final strength of the bonded joint is not to be expected. However, if the wet bonding time is exceeded and the fixing of the adherends is carried out afterwards, the adhesive strength will be weakened (exception: contact adhesives). Thus, the open assembly time comprises the minimum drying time.

In this context, the term "wet adhesive" requires a special explanation. In processing, it serves the differentiation between contact adhesives and solvent-based adhesives and describes the still (partially) liquid state of the adhesive prior to the fixing of the adherends. This term is not a common designation of adhesives in the broader sense.

Figure 5.3 describes the setting principle of solvent-based adhesives.

The following criteria are important for the selection and the processing of solvent-based adhesives:

- *Porosity of adherends:* Depending on the pore size, the applied adhesive may penetrate into the material, leaving an insufficiently thick adhesive layer. A remedy would be a second adhesive application (on each adherend) after a short period of time, or an adhesive of higher viscosity.

- *Adherend or ambient temperature:* The higher the temperature, the faster the evaporation of the solvents; this again leads to a reduction of the open assembly time.

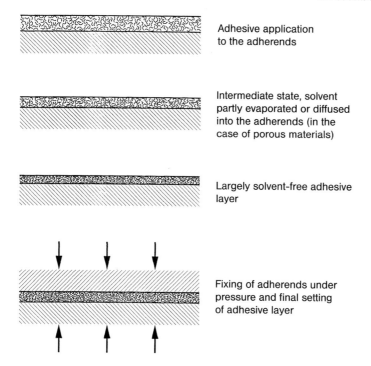

Figure 5.3 Setting principle of solvent-based adhesives.

- *Amount of adhesive applied:* The thicker the adhesive layer, the more solvents must evaporate. This may result in an extension of the open assembly time.
- *Application of surface pressure:* When processing solvent-based adhesives, it is in any case recommendable to apply equally distributed pressure to the adhesive layer after the fixing of the adherends. This causes the polymer molecules existing on both adherends to get "jammed" or "entangled", which increases the cohesion strength of the adhesive layer.

Solvent-based adhesives are particularly suitable for the bonding of porous materials, such as paper, cardboards, wood, cork, leather, textiles and foams.

▶ **Supplementary Information:**

- For the application of solvent-based adhesives to impermeable adherends (metals, glasses) the compliance with the maximum drying time has to be observed. If the fixing of the adherends occurs too early, too much solvent remains in the glueline, affecting the strength of the adhesive layer considerably. Since the complete evaporation of the remaining solvents is only possible over the edges of the bonded joint, very long setting times are to be expected.
- In the case of permeable, respectively, porous materials (papers, cardboards, wood, etc.) the fixing of the adherends is already possible when the minimum

drying time (Figure 5.2) has been reached. Then, existing solvent residues evaporate through the adherends.
- Since the organic solvents used are combustible, special care has to be taken to keep these adhesives away from ignition sources during their processing. Furthermore, it has to be mentioned that vapors of organic solvents are usually heavier than air and sink to the bottom during evaporation, where they can "creep" so that even more distant ignition sources may cause an ignition. In general, bonding with solvent-based adhesives should be carried out under suitable exhaust units. For ecological reasons, the tendency in adhesive development has already gone to solvent-free or at least low-solvent adhesive formulations.
- When bonding *polystyrene foam*, care has to be taken that only so-called "polystyrene-neutral" adhesives are applied. The reason is the ability of polar solvents (e.g., chloroform, acetone) to dissolve the polystyrene and thus to destroy the foam structure. For such applications, special polystyrene adhesives are available.
- For plastics soluble in solvents such as Plexiglas, polycarbonate, polyvinyl chloride, special solvent-based adhesives are available (Section 9.2.5).
- For large area applications, solvent-based adhesives are also sold as "spraying adhesives" in cans.

5.3
Contact Adhesives

Contact adhesives are characterized by the fact that the solvents contained in the applied adhesive have to evaporate completely, prior to the fixing of the adherends (depending on the adhesive quantity applied, 15–20 min), until the adhesive layer feels "contact-dry" to touch. Thus, the maximum drying time according to Figure 5.2 is exceeded. Afterwards, the adherends are assembled at the highest possible contact pressure so that the formation of a bonded joint with relatively high strength will be achieved in a very short time. Apart from the mutual penetration ("jamming, entangling") of the polymer molecules, even the development of crystal structures in the adhesive layer contributes much to this strength.

Contact adhesives have two forms of application:

- *One-sided bonding (wet bonding):* In this procedure, the adhesive is applied to *one* adherend only. This is recommendable for bonding of solvent-permeable, respectively, absorbent materials (leather, textiles, wood products). In this case, complete solvent evaporation is not required.
- *Double-sided bonding:* This refers to the actual "contact bonding" that has to be applied whenever solvent-impermeable or dense materials (metals, glass, plastics) are to be bonded or when the bonded joint is required to show very high initial strength. Absorptive materials may require a second adhesive application.

Important polymers for contact adhesives are natural and synthetic rubber types, in particular polychloropene rubber and polyurethane polymers.

A typical example of contact bonding is the patching up of a rubber tube with a "rubber solution". After evaporation of the solvent, the rubber polymers in the rubber solution combine with themselves under pressure and with the parts of the rubber surface areas swollen by solvents.

▶ **Supplementary Information:**

- Due to the development of the adhesive layer described above, high contact pressure is more important for the strength of a contact bonding than long contact time.
- The possibility to bond solvent-impermeable materials, such as metals, glass, plastic-coated boards and so on with contact adhesives is also advantageous – in contrast to the solvent-based adhesives described in Section 5.2 – according to the double-sided method.
- Since contact adhesives form flexible adhesive layers that are deformable to a certain extent and are thus able to compensate for material stresses, they are particularly suitable for materials such as leather, rubber, sole material (shoe industry) and so on.
- Due to the long "open assembly time" inherent in the contact adhesives (Section 5.2) they are also applied to bonding of large area surfaces, for example, in wood processing for the application of veneers.
- When fixing the adherends, it has to be considered that a later adjustment is not possible. It is recommendable to use a fixing device (edge, or the like, to align the adherends with).
- In addition to the one-component contact adhesives described above, there are also two-component systems with isocyanate compounds as hardener component as well as solvent-free formulations.

5.4
Dispersion Adhesives

A distinctive feature of dispersion adhesives, in contrast to solvent-based adhesives described in Section 5.2, is that water is used as an incombustible "solvent". This bears an advantage regarding possible processing risks and environmentally relevant regulations. Therefore, adhesive and raw material manufacturers made considerable research and development efforts to replace solvent-based adhesives by dispersion adhesives.

What are dispersions? From a physical point of view, they are so-called multiphase systems in which a "phase" describes the state of a certain substance.

The phases include:
- most materials in *solid phase* (metals, plastics, glasses, minerals);
- water, oils, solvents in *liquid phase;*
- gases known to us in *gas phase* (oxygen, nitrogen, but also water vapour).

In the case of dispersions (from Latin dispergere = fine distribution) a solid phase has dispersed into the liquid phase (water as dispersing agent) owing to the polymer particles in diameter ranges of 10^{-4}–10^{-5} cm (ten-thousandth to hundred-thousandth cm). The smallness of the particles and special additives (stabilizers, emulsifiers) prevent them from deposition. The solid content ranges between 40–70%. The setting mechanism for the formation of the adhesive layer is initiated by the removal of the liquid phase. This may occur by

- evaporation of water and/or
- penetration of water into the adherends.

Since the latter possibility is only given in case of porous surfaces, dispersion adhesives are mainly used for materials that are capable of binding the water of the liquid adhesive layer by "absorption". The remaining polymer particles with their inherent tack fuse to an adhesive layer (film formation). The open assembly time (Section 5.2) in the processing of dispersion adhesives is strongly influenced by the moisture content of the adherends and the relative humidity. The open assembly time increases with higher humidity.

For dispersion adhesives, the most important polymers are polyvinyl acetate, acrylate, rubbers, polyurethane and polychloroprene.

Among the preferably bonded materials there are, *inter alia*, wood and wood products such as hardwood and softwood, chipboards, hardboards, plywood, veneers, tongue and groove joints, dovetails and tenon and mortise. Due to their being solvent-free, dispersion adhesives are also suitable for the bonding of polystyrenes.

▶ **Supplementary Information:**

- Since the water content of dispersion adhesives must be absorbed by the adherends during setting, the setting time depends mainly on the moisture content, above all in the case of wood (favorable range 8–10%) (Section 9.5).
- When choosing dispersion adhesives suitable for wood, the later stress on the bonded joint, especially moisture, has to be taken into account (Section 9.5).
- The dispersion adhesive is usually applied to one side, in the case of very rough cutting edges and hard woods also on both sides. Residues from wood processing must be previously removed.
- Bond the adherends as long as the adhesive is wet, then fix the bonded joint under pressure.
- Protect dispersion adhesives from frost, since after thawing they are no longer usable due to the distruction of the dispersion.
- Metals, glasses and other impermeable materials are not bondable by dispersion adhesives.

5.5
Plastisols

In their processable mixture, these one-component products, also belonging to physically setting adhesives, consist of two components: PVC (polyvinyl chloride-) particles and plasticizers (Section 9.2.9). The solid PVC particles are disperged in the high-viscosity plasticizer. The adhesive layer formation occurs by heating (120–180 °C), when the thermoplastic PVC swells and is thus able to absorb the plasticizer (no chemical reaction!). This process is called a sol-gel process. The formerly two-phase system (sol) is turned into a single-phase system (gel) by the inclusion of the plasticizer.

They are typically applied as adhesive sealants in bodywork manufacturing (fold bonding and flange bonding, vibration insulation, corrosion protection) and as sealants in bottle and glass caps. For environmental reasons (hydrochloric acid separation in the case of thermal disposal), PVC plastisols are increasingly being replaced by acrylate-plastisols and epoxy systems.

5.6
Pressure-Sensitive Adhesives, Adhesive Tapes

Pressure-sensitive adhesives are the essential components of adhesive tapes and labels. They are polymers with permanent tack, usually applied on substrates (plastic/metal films, siliconized papers). To enhance their tack, compounds with high inherent tack are added, for example, resins, plasticizers. Pressure-sensitive adhesives reach their adhesion on the material to be bonded by contact pressure, from which the term *pressure-sensitive adhesive (PSA)* derives. Apart from electron radiation, also UV-radiation curing described in Section 4.3.2 is applied in adhesive tape manufacturing. The monomer molecules to be polymerized are applied, in liquid form, to the substrates to be coated by rolling and are continuously cured to a polymer layer within seconds under a UV-radiation source. Depending on their composition, predetermined adhesion values can be adjusted. The adhesive tapes can be subsumed under the systems shown in Figure 5.4:

- Transfer adhesive tapes: Adhesive films consisting 100% of the respective pressure-sensitive adhesive polymer (mostly acrylate). For their processing, they are applied on a separable substrate.
- One-side adhesive tapes: Adhesive tapes with a substrate to one side of which the adhesive layer is applied and bonded.
- Two-side adhesive tapes: Adhesive tapes with a substrate of which the adhesive layer is applied to both sides.
- Foamed adhesive tapes: Adhesive tapes without a substrate, with the total system consisting of a pressure-sensitive adhesive polymer with two-sided adhesive properties available in foamed and closed-cell structure. They are not to be confused with one or two-side adhesive tapes with foam-structure substrates. In

5 Physically Setting Adhesives

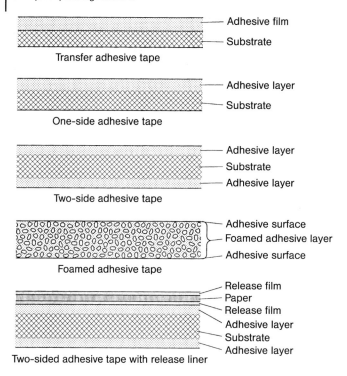

Figure 5.4 Structure of adhesive tapes.

this case, the foam elastically evens out the glueline – possibly with the drawback of remaining stresses – the thin pressure-sensitive adhesive film, however, can only bridge minor surface roughnesses.

In order to avoid the sticking together of the two-sided and foamed adhesive tapes when rolling them up, siliconized release liners are required. The specifically formulated release layers enable the residue-free stripping of the adhesive tape. The same release liners serve the temporary fixation of adhesive labels.

Depending on the formulation, permanent pressure-sensitive adhesives (e.g., for car tax vignettes, price-labels) or removable pressure-sensitive adhesives are available. In the past, so-called "repositionable" pressure-sensitive adhesives have gained special importance, which are mainly applied to papers used as multi-adhesive memo sheets.

▶ **Supplementary Information:**

- When applying adhesive tapes, it has to be considered that a correction of the adherend position after fixing is no longer possible.
- Adhesive tapes are also suitable as fixing aids for the manufacturing of bonded joints to secure adherends against moving (in the case of heat-curing adhesives, however, applicable only to a limited extent).

- **Tip:** When bonding with pressure-sensitive adhesives, first degrease the surface of the material, then warm the adhesive area and the pressure-sensitive adhesive layer with hot air (hair-dryer) and fix it immediately under pressure. This procedure renders the adhesive layer "more flexible" and enables better levelling out of the surface structures. Moreover, the contact surface will be extended.

5.7
Adhesive Strips

In addition to the adhesive tapes mentioned in Section 5.6, *adhesive strips* are paper strips mostly consisting of kraft paper coated with an adhesive layer that is water or heat activateable. The adhesive layer, generally called "gumming", consists of a water-activateable adhesive, initially applied to the substrate in liquid form and then dried. The adhesive properties of the gummed layer develop due to the humidification of the adhesive strip in processing. Mainly animal-based products (animal glue) and plant-based products (starch glue, dextrine glue) are used as basic materials for the adhesive. In the case of heat-activateable adhesive strips, hot-melt adhesive coatings activated by means of hot air or infrared radiation are applied to the paper strips.

5.8
Glue Sticks

It is common practice to apply adhesives without solvents and/or heat supply by simply rubbing them onto the material to be bonded. In the glue sticks, solid adhesive formulations are rod-like arranged in a reclosable capsule. When rubbing them off, for example on a paper surface, they leave a sticky film. Glue sticks usually contain water-soluble or water-dispersing polymers with adhesive properties embedded in a form-giving structural substance, a so-called "soap gel". This substance is structured in a way that it will be destroyed under mechanical stress, in this case by shear forces when rubbed off, and releases the adhesive components. Depending on the desired purpose, glue sticks with permanent or removable properties are available.

5.9
Adhesives Based on Natural Raw Materials

In contrast to the "young" adhesives on synthetic bases, adhesives deriving from natural products have partly been known for millennia. The essential differences compared to reactive adhesives are the partially poor ageing stabilities in humid atmosphere as well as the low bonding strengths. They are not used for highly strained bonded joints of metals, plastics, glasses and similar. However, they record a big market share – partly in modified form – in bonding of paper and cardboard products (wallpaper, packaging, labels of glass and plastic containers) and in wood processing. For special demands, for example, waterproofness, adhesives combined with the corresponding synthetic resins are available.

In the application field of natural adhesives, the traditional terms "paste" instead of "adhesive" or "glueing" instead of "bonding" are still in use.

According to the definition:

- *Glue:* is an adhesive consisting of animal and/or plant raw materials (as the case may be, even mixed with synthetic portions) as well as of water as solvent.
- *Paste:* is an adhesive in the form of an aqueous swelling product that, in contrast to glues, forms a nonropey, high-viscosity mass already at low basic material concentrations.

The development of the adhesive layer follows the principle of physical setting (Section 2.2.2) on simultaneous evaporation or absorption of the water by the (porous) adherends.

A distinction is made between products on:

- *Animal base:* skin, bones, leather, fish and casein glue (most important glue for bottle labelling), glutin hot-melt glue.

- *Plant base:* starch, dextrin, cellulose glue, gum arabic.

The basic substance of wallpaper paste is cellulose (wood component) that, with regard to the required processing properties (water solubility, strength, removability as the case may be) is chemically modified.

5.10
Adhesives on an Inorganic Basis

This adhesive group is used in the case of very high thermal resistance of the glueline. In general, it is to be noticed that inorganic compounds show a considerably higher form of stability and thermal resistance than organic compounds (Section 2.2.5), due to the chemical-bonding conditions in the molecules. In particular, the "waterglass", especially sodium silicate has to be mentioned that is mainly used for waterproof bonding of papers and cardboards. It consists of an aqueous, colloidal solvent of sodium silicate. Setting occurs by evaporation of water through the porous materials and the ensuing formation of silicic acid structures. Formulations on the basis of inorganic products (silicates, metal oxides, borates) are also used for glass plunger/cap bonding of light bulbs and halogen lamps.

Complementary literature to Chapter 5:
[B1, B2, B3, B5, S2].

6
Adhesive Forces in Bonded Joints

6.1
Adhesive Forces Between Adhesive Layer and Adherend (Adhesion)

A frequently asked question refers to the reasons for the development of the adhesion of adhesives/adhesive layers on surfaces. One answer sees the reason for this phenomenon in the existence of rough surfaces in which the adhesive layer gets "interlocked", and is thus "positively connected" to the adherend, in the sense of the description given in Section 1.1 (Figure 6.1).

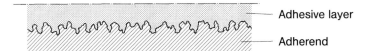

— Adhesive layer

— Adherend

Figure 6.1 Positively bonded joints of adhesive layer and adherend.

This positive or "mechanical interlocking" (it is also talked of "mechanical adhesion") is indeed a possibility of joining adhesive layer and adherend. It occurs preferably in the case of very rough and/or porous surfaces, for example, of papers, cardboards, wood, ceramics or plastic foams. This concept fails in the case of smooth surfaces, however, surfaces we call "smooth" can actually show a "mountainous structure" under the microscope. But such fine roughnesses can hardly contribute to sufficient mechanical interlocking. Thus another possibility must exist which enables the adhesive layer and the adherend to be joined firmly and permanently.

Before these relationships are discussed *adhesion*, as a new term, has to be explained. The word is of Latin origin (*adhaesio, adhaerere*) and means something like "to stick to something, to adhere". The term "adhesion" is very frequently used and belongs to the standard terms in bonding technology.

How to explain adhesion? We know a lot of examples of everyday life in which substances adhere to other substances, for example, fine dust particles on window panes or plastic rails, water drops on a vertical surface.

Glass sheets or polished metal surfaces may adhere to each other so that they can hardly be separated. The reason for this behavior is the internal structure of

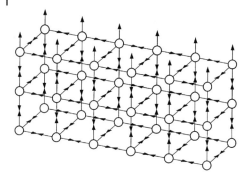

Figure 6.2 Adhesive forces on adherend surfaces.

the materials. All substances known to us consist of atoms and molecules with their cohesion being based on electric forces. Inside the material these forces are equally distributed between the atoms and/or the molecules. In the surface area, the atoms and molecules no longer have equal "neighbors" they can exchange electrical forces with, therefore they act into the surrounding atmosphere. They are capable of binding other substances, for example, dust particles or water drops. "Adhesion" takes place. This kind of adhesive forces are also called *dipole forces* (Figure 6.2).

Such force actions are developed in a similar way by polymer molecules, which then fuse to form a strong bond with those of the adherend surface as shown in Figure 6.3.

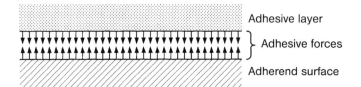

Figure 6.3 Adhesive forces between adherend surface and adhesive layer.

Thus, such adhesive forces are the basis for the fact that a bond, consisting of adherends and adhesive layer, will last. Since they develop between the individual molecules or atoms, they are also called *intermolecular forces*. The distances over which adhesive forces can act are very short, they range in the order of 10^{-5} (one hundred thousandth) millimeters. So, it is to be explained that very smooth, finely polished surfaces can still have a certain adhesion to each other. Thus, part 1 in Figure 6.4 (left) is able to lift part 2 (without being connected to part 1 by an intermediate layer) from the base area. In the depiction on the right this is not possible due to the surface roughness.

This consideration makes clear that adhesive forces can only be effective, if no other material, for example, the mentioned dust particles or moisture layers, has bonded with the adherend surfaces before the adhesive will be applied. Then, no

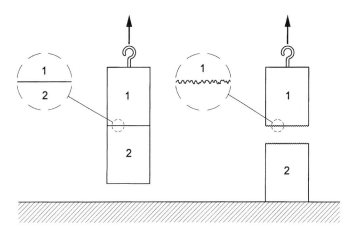

Figure 6.4 Development of adhesive forces at ideal smooth and rough surfaces.

or only a small number of adhesive forces are available for sufficient "adhesion" of the adhesive layer to the surface. So, if processing instructions of the adhesive manufacturers indicate: "The parts to be bonded should be dry and free of dust and fat", this advice is based on the previously described relationships. Section 7.1 is dedicated to a more detailed discussion of this relationship for surface treatment as an essential step in the production of bonded joints.

6.2
Wetting

On the assumption of a clean and, respectively, pretreated surface, the next step will be the application of the adhesive. It has to be ensured, however, that in areas with adhesive forces emanating from the adherend surface, the adhesive molecules are really able to get closer. Only then can the adhesive distribute itself on the surface, that is, wet the surface despite a, more or less, existing roughness. Furthermore, sufficient flowability of the adhesive is important. A complete and equal wetting of the surface to be bonded is therefore an indispensible prerequisite for the production of a strong bonded joint. Figure 6.5 demonstrates the difference between a low-viscosity and high-viscosity adhesive.

Depending on the adhesive viscosity and the wettability of a surface, liquid drops applied to a surface take on different forms. Characteristic, in this connection, is the contact angle α that develops between the liquid adhesive and the adherend surface. The smaller the angle, the better the wetting. Good wetting is talked of, if the values of α are below 30°. If an adhesive (or a liquid in general) spreads spontaneously, that is, without external influences (like rolling, daubing, knife coating), and evenly on a surface, the wetting behavior is very good. In this case one talks of *spreading* (Figure 6.6). The case of $\alpha \sim 180°$ (spherical shape), for example, is known for mercury drops.

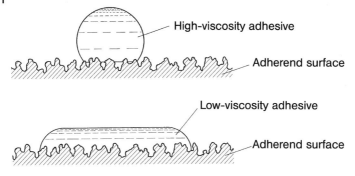

Figure 6.5 Wetting behavior of a high-viscosity and a low-viscosity adhesive.

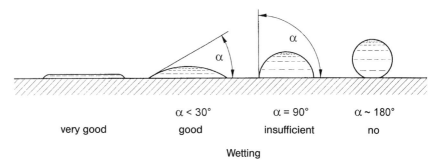

Figure 6.6 Wetting forms of liquids on surfaces.

6.3
Surface Tension

The wetting behavior of a liquid on a surface not only depends on its viscosity, as described in Section 6.2, but also decisively on its surface tension. This term derives from the idea that a water drop, for example, is prevented from decomposing by an invisible film it is "embraced" by. The real reason for this appearance can be deduced from Figure 6.7.

While inside a liquid, from all directions the same attractive forces act on the water molecules of the given example, at the interface of the water drop to the ambient air, such forces are not balanced. Thus, there is a force F directed to the inside of the drop, which tries to draw the water molecules away from the surface into the only inside of the drop. Consequently, the drop aims to reduce its surface, which results in the formation of the spherical/drop form. (The sphere is the geometrical form with the smallest ratio of volume and surface area.)

Apart from liquids, also solids, such as metals, glasses and plastics have surface tension. Due to the stiffness of these materials, it is invisible to the eye, but metrologically determinable. Thus, with the application of the adhesive, two partners with different surface tensions are joined – depending on the material of the adherend and the adhesive.

Force effect on the interface

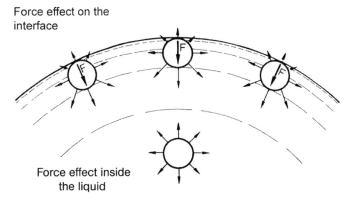

Force effect inside the liquid

Figure 6.7 Surface tension of liquids.

According to the laws of thermodynamics, the difference of surface tension between adherend and adhesive is decisive for the wettability of the system. The surface tension values are given in mN/m (milli Newton per meter) in the following order of magnitude:

Metals	1000–3000
Glasses	250–350
Water	72.8
Plastics	20–60
Adhesives	30–50

From these values it is clear that the respective difference for metals compared to adhesives is rather large, while for plastics compared to adhesives it is rather small. In practice, this means:

- After the appropriate preparation, metals have good to excellent bonding qualities (owing to their "noble" character, however, noble metals like gold, silver, platinum, etc., are an exception).
- Plastics require special conditions to be bonded. The fundamentals are described in Section 9.2.

6.4
Adhesive Forces Inside an Adhesive Layer (Cohesion)

Solid bonded joints not only require adhesive forces sufficient in quantity and strength, but also a certain strength of the adhesive layer between the adherends. This requirement can be pointed up on the basis of Figure 6.4. If some drops of a runny oil are applied between the two rough (right) adherends – after prior degreasing to enable good wetting of the surfaces – the lower adherend can be lifted up from the base area together with the upper adherend despite the rough

6 Adhesive Forces in Bonded Joints

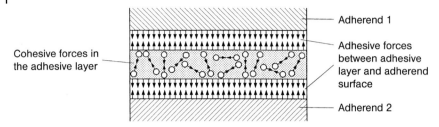

Figure 6.8 Adhesive and cohesive forces in a bonded joint.

surface, thus both parts "stick together" due to the adhesive forces of the oil film acting almost like an adhesive. However, they can easily be shifted against each other, since the oil layer is liquid and not solid. Although adhesives are liquid at the moment of their application, they solidify to a solid intermediate layer that does not allow shifting of the adherends. Thus, compared to a liquid, their "internal forces" causing the cohesion are much stronger. With reference to the Latin word *cohaerere* = holding, stickring together, the forces existing in an adhesive layer are called *cohesive forces*. Cohesive forces act in all solid and liquid substances. The stronger they are, the better the dimensional stability of a substance.

Therefore, the production of an adhesive bond requires that the adhesive layer shows equally developed cohesive strength all over the glueline. Wrong mixing ratios of the components or air bubbles creeping in during the mixing procedure may have interfering effects. Another reason for insufficient cohesive strengths may be the non-observance of the required curing time or temperature.

A summarizing description of the forces acting in a bonded joint is given in Figure 6.8.

Thus, *adhesion* is the sticking together of same or different substances, *cohesion* is the inner strength of a substance, in this case of the adhesive layer. The sphere of action of adhesive forces is defined as a boundary layer, see also Figure 1.3.

Complementary literature to Chapter 6:
[C3, I1, L2, M3, M4, M5, P1, P4, W1, W3].

7
Production of Bonded Joints

Following the description of the fundamentals regarding structure and different kinds of adhesives as well as bonding relationships effective in bonded joints, now the most important process steps will be depicted. Here, two groups are to be subdivided (Figure 7.1):

- Processes serving the development of the adhesive forces. They include surface treatment of the adherend and adhesive application.
- Processes defining the cohesive strength of the adhesive layer. In this case, the conditions in respect of time, temperature and pressure during adhesive curing have to be taken into account.

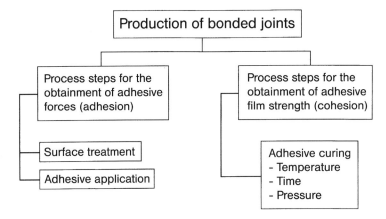

Figure 7.1 Process steps for the production of bonded joints.

Applied Adhesive Bonding: A Practical Guide for Flawless Results. Gerd Habenicht
Copyright © 2009 WILEY-VCH Verlag GmbH & Co. KGaA, Weinheim
ISBN: 978-3-527-32014-1

7.1 Surface Treatment

Depending on the given conditions, the three different process types shown in Figure 7.2 can be applied:

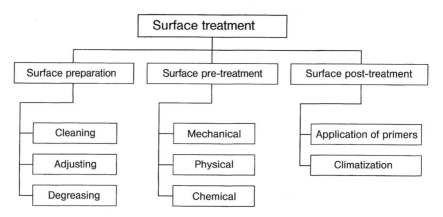

Figure 7.2 Processes of surface treatment.

7.1.1 Surface Preparation

7.1.1.1 Cleaning
Cleaning of the adherend surfaces serves the removal of adhesive solid layers like dirt, rust, tinder, paint, lacquers, and so on. Mechanical cleaning by means of grinding or brushing is preferred. Cleaning is a prerequisite for the aspired strength of a bonded joint even with low stress, since impurity layers are regarded as the basis for glueline failures.

7.1.1.2 Adjusting
This work step is mainly required for the obtainment of equal adhesive layers. Here, particularly in the case of small adhesive surfaces as used for test purposes, it is necessary to remove the burr at the test specimen. In the case of larger adhesive surfaces, the rectification of the adherends is a prerequisite for parallel gluelines.

7.1.1.3 Degreasing
Degreasing may occur by means of organic solvents or hot distilled water (approximately 60–80 °C) added with liquid cleaning agents (approximately 1–3%). Attention should be paid since the cleansing agents may contain small fractions of silicone compounds, which will complicate wetting when remaining on a surface. Degreasing is one of the most important prerequisites for perfect

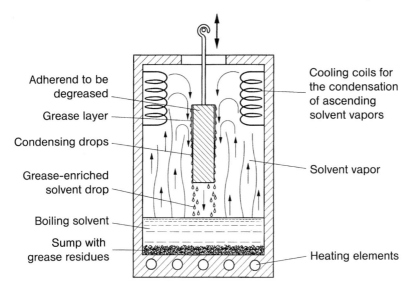

Figure 7.3 Principle of vapor degreasing.

wetting, therefore it should be carried out in any case irrespective of whether or not further surface pretreatment will follow.

The degreasing method to be applied depends on the number of items to be degreased, on the form of the adherends and the required degree of degreasing:

- The simplest possibility of degreasing adherends is wiping them with solvent-soaked, nonfuzzy cloth or tissues as well as immersing them. However, both procedures bear the disadvantage of an uncontrolled degree of degreasing through uneven wiping or possible fat accumulations in the solvent. However, there are no alternatives to these two possibilities for adhesive bonding on the nonindustrial level or for do-it-yourself applications.
- For technical applications aimed at the obtainment of utmost and equal freedom from grease, *vapor degreasing* is applied. In this procedure the adherends are immersed in the solvent vapor phase and heated, depending on the boiling point of the utilized solvent. Due to the solvent condensation at the initially cold adherends, the grease particles are "washed up", with the advantage that regreasing through grease accumulated in the "sump" of the degreasing facility is practically impossible (Figure 7.3).

7.1.1.4 Degreasing Agents

The good grease solvents trichloroethylene or perchloroethylene, mainly used in the past, were increasingly recognized as environment killers and as jointly responsible for the degradation of the ozone layer in recent years and are therefore no longer applied. Alternatively, acetone, methylethylketone (MEK), ethylacetate or also methyl- and isopropylalcohol show good degreasing properties. Benzene and petrolether in nonpurified form are not recommendable, because they could

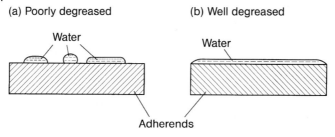

Figure 7.4 Poor (drop formation (a)) and perfect (equal distribution (b)) degreasing.

contain paraffins that remain on the surface after evaporation and make the wetting more difficult.

A simple method to determine the degree of degreasing is wetting the surface with demineralized water by immersing or dropping. If a surface is wettable with water, it will be wettable by adhesives, too, since the latter show better wetting properties compared to water (Figure 7.4).

▶ **Supplementary Information:**

- Due to their good fat solubility, organic solvents also affect the natural fat layers of the skin, so in the case of permanent contact skin damages are possible. As preventive measures, the wearing of rubber gloves and the use of skin-protection cream are strongly recommended.
- Due to the possible fat transfer from the skin to the surfaces, the latter must by no means get into contact with the naked hand after degreasing.
- Caution: Organic solvents are combustible! The used solvent-drenched cloths must be kept away from flames in closed containers and have to be adequately disposed of (Section 7.5.1).

7.1.2
Surface Pretreatment

Following the surface preparation, it is the task of surface pretreatment to generate the adhesive forces on the adherend surfaces required for the development of a strong bonded joint. Since almost all materials interesting for bonding have the property to cover the surfaces with impurity layers (oxides, rust, dust, greases), those layers have to be completely removed prior to adhesive application, since otherwise failures in the development of the adhesive forces will occur (Figure 7.5).

Figure 7.5 Impairment of the adhesive forces by surface contamination.

7.1.2.1 Mechanical Surface Pretreatment

Grinding, brushing, sanding or blasting are the most important methods. Prior degreasing is required at any rate, since otherwise probable grease residues could be spread on the surface or even pressed into fine pores or other depressions.

Grinding, brushing and sanding are characterized by low dust load in comparison to blasting, however, the evenness of the pretreatment leaves much to be desired. The grinding or brushing effect can be enhanced by repeating the process at an angle of 90° (crossgrinding). For the pretreatment of larger surfaces, pad sanders and band grinders are available. More effective than grinding and brushing is sanding (grit blasting) with shot sold in different kinds and forms (aluminum oxide abrasive, steel grit, glass pearls). Considering the costs, especially in the case of long-term stress of bonded joints, blasting can be regarded as the ideal pretreatment method. Depending on jet pressure and abrasive grain diameter, a more or less rugged surface develops, as shown for steel in Figure 7.6.

The roughnesses achievable during blasting depend on the jet pressure and the grain size of the shot; they range between 50 and 100 µm (1 µm = 1 micrometer = 0.001 mm).

If the dimensions of the surfaces to be prepared allow, blasting is carried out in closed steel cabins equipped with a collecting and recirculation device for the repeated application of the shot. Since in these cabins the emission of dust particles cannot be excluded, it is recommendable at any rate to install them in rooms separated from bonding work. For larger surfaces so-called back-suction jet systems are recommendable, where the shot will be fed back into the jet circuit by means of a suction device concentrically arranged around the outlet nozzle.

Due to the abrasive grains striking the adherend surfaces with high energy by means of compressed air, surface densification with ensuing development of tension may occur, entailing a deflection, especially in the case of thin adherends (sheet metal up to 2 mm thick). This phenomenon may be avoided by clamping

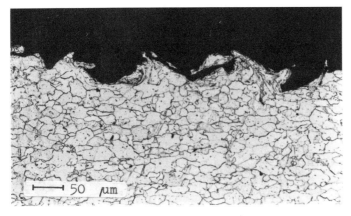

Figure 7.6 Blasted steel surface.
Abrasive grain: corundum; grain diameter: 0.5–1 mm;
jet pressure: 0.8 MPa; nozzle distance: approximately 100 mm.

the sheet metal onto a thick, inflexible base. The deflection can be reversed by blasting of the adherend's backside. This drawback does not occur in grinding and brushing.

As the compressed air required for blasting is generated in compressors, the presence of small quantities of oil cannot be excluded, which remain on the surface after blasting. For this reason, ensuing degreasing is essential, with the additional advantage that possibly existing shot residues in the surface finish will also be removed.

As a conclusion, it can be said that measured against the required time and effort, a surface pretreatment with the stages

Degreasing – Blasting – Degreasing

provides the optimal preconditions for the production of bondings with good long-term resistances.

A special variant of blasting is the "Saco" method (DELO-Industrieklebstoffe GmbH, 86949 Windach, Germany), where the shot not only acts mechanically on the surface, but due to the special composition, it also leads to chemical changes of the surface, which may result in further improvements of the behavior of the bonded joints.

▶ **Supplementary Information:**

- The use of a sanding block made of wood or cork covered with abrasive paper, if required fixed with an adhesive tape, has proven its worth.
- Good cleaning results are also reached with so-called wired sponges, as used in households.

7.1.2.2 Physical and Chemical Surface Pretreatment

Grinding, brushing or sanding (with the exception of the above-mentioned Saco method) do not cause chemical modifications of the material's surface. A clean surface results with a characteristic structure corresponding to the composition of the material, as shown in Figure 7.6. Therefore, physical and chemical pretreatment methods are aimed at the chemical modification of the surfaces. Thus, on the one hand it is possible to further enhance the adhesive forces for extremely high demands on bonded joints, and on the other hand, to make poorly bondable material (e.g., plastics) bondable at all. Since physical methods are mainly used in bonding of plastics, they are described in Section 9.2.4.

All chemical methods bear the disadvantage of requiring aggressive chemicals strongly detrimental to health. Their application is subjected to legal obligations, and thus to strong safety regulations, and requires high expenditure for the disposal of the chemicals after their use. For this reason in industry they are only applied in exceptional cases, where a particularly long life of the bonded joints, which are at the same time exposed to high stress, such as corrosion, has to be guaranteed. An example of this is the aerospace industry with service lives of aircrafts of up to 30 years.

7.1.2.3 Pickling

The group of chemical surface pretreatment methods also includes pickling. Here, thinned acids are applied, which remove layers on the metal surfaces via chemical reactions resulting in metallically clean surfaces. The respective application regulations apply, too.

7.1.2.4 Surface Layers and Creep Corrosion

For the application of the explained surface pretreatment methods and their effects, the structure of surface layers typical for metal materials will be described below (Figure 7.7).

Initially, there is a "reaction layer" on the basic material, the surface of which can be selectively modified with regard to its mechanical and metallurgical properties during its manufacturing (by cold deformation when temper rolling). The name derives from the property of many metal materials to chemically react with the components oxygen, water vapor, etc. existing in the air. Characteristic examples of this is the corrosion of iron or the tarnishing of silver. These layers, chemically modified compared to the basic material, show more or less adhesion on the basic material according to the conditions given during their formation. It is known that rust layers can spall from the surface. Thus, such naturally grown surface-reaction layers cannot be a precondition for the formation of strong adhesive layers. They have to be removed prior to the application of the adhesive, either mechanically by grinding, brushing or sanding, or even by chemical pickling. These layers cannot be removed by degreasing!

If required, reaction layers with very good adhesive properties will afterwards be applied to the clean surface, by targeted chemical reactions under exactly defined conditions. Then, bonding on the surface is possible. For this purpose, chemical and electrochemical prepreparation methods are used that, however, require a lot of time and effort, as already explained.

The adsorption and contamination layers shown in Figure 7.7 are removed during the mechanical surface pretreatment. Such layers, for example, adsorbed moisture or dusts, quickly develop again on clean surfaces. For this reason, the time between the surface pretreatment and the application of the adhesive should

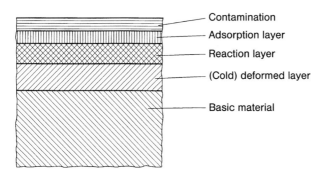

Figure 7.7 Surface layers of metal materials.

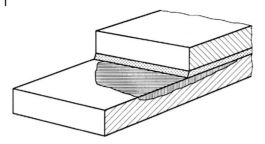

Figure 7.8 Creep corrosion of adhesive layers.

be as short as possible. Dust and adsorption layers can be removed effectively again by degreasing.

Finally, an important difference between mechanical and chemical surface pretreatment methods should be pointed. Concerning their effects, the former are only limited to the adhesive surface area, that is, they are not able to protect the neighboring areas of the adhesive surface against possible stress from the ambience.

What is the use of an optimum surface treatment if creep corrosion affects the adhesive layer from outside the glueline, which results in the destruction of the bonded joint? This process is shown in Figure 7.8.

7.1.3
Surface Post-Treatment

A surface post-treatment does take place, if the adhesive properties of a surface shall be further enhanced or bonded joints are exposed to particularly high stress, for example, by moisture and corrosion. Here, it has to be considered that longer periods under frequently uncontrolled storage conditions may occur between production and processing of the material to be bonded.

7.1.3.1 Primer

Usually, primers for adhesive applications are mainly applied to the material directly after manufacturing, however, application is also common after the pretreatment of the surface. Primers consist of solutions of polymers, partly even of reactive monomers in organic solvents that, in their composition, are related to the adhesives and applied to the surface in a thin layer (up to approximately 5 g/m^2) by immersing, brushing or rolling. After drying, if required even at elevated temperatures, the primer layers create a very good prerequisite for the production of stronger and more resistant bonded joints. In this context, it has to be emphasized that primer and adhesive have to be absolutely adjusted to each other, and only the primer the adhesive manufacturer prescribes for a certain adhesive may be used.

7.1.3.2 Climatization

For the sake of completeness, climatization deserves attention within the context of the surface post-treatment. With fluctuation of temperature and humidity, water condensation on the adherends is possibly limiting the adhesive properties.

Thus, the described possibilities of surface post-treatment serve two aims:

- to maintain or improve the adhesion conditions resulting from the respective pretreatments and
- to avoid the uncontrolled modification of a surface even after completed bonding, as, for example, in the case of corrosion creep. For special conditions, it is therefore necessary also to protect the areas surrounding the bonded joint against external influences.

> *In conclusion, the creation of defined surface properties for reproducible bonding results is the decisive challenge of surface treatment.*

7.2 Adhesive Processing

7.2.1 Adhesive Preparation

Even if in most cases the adhesive supplied by the manufacturer can be assumed to be in a processable state, it cannot be excluded that, prior to application, certain preparations are required. This work step may comprise the following measures:

7.2.1.1 Viscosity Adjustment

This is important in the case of solvent-based adhesives and dispersions. In order to achieve equally thick adhesive layers – especially in roller and spray applications – the adhesive must be applied with the prescribed viscosity. When storing solvent-containing adhesives, slightly leaking containers may cause evaporation of the solvent and thus an increase in viscosity. Only diluting agents prescribed by the adhesive manufacturer may be used for dilution (Attention when diluting: fire and explosion hazard!). Definition and dimensioning of viscosity, see *Common Terms* in Chapter 13.

7.2.1.2 Homogenization

If adhesives are applied with *fillers* added to achieve special processing properties (e.g., to be able to produce thicker adhesive layers), such fillers can deposit and must then be stirred to enable equal distribution. Care has to be taken that no air is stirred into the adhesive, since this will lead to pores in the adhesive layer during the ensuing adhesive application and curing (stirring in the vacuum, if required).

7.2.1.3 Climatization

The *climatization* of the adhesive to the processing temperature is important in cold or hot seasons and in the case of its storing in nonconditioned rooms to ensure constant processing viscosities. In the case of watery adhesive dispersions, the risk of freezing has to be pointed out in this context, which results in a destruction of the dispersion and thus in uselessness – even after thawing. Independent of the kind of adhesive application, it is sensible to adhere to the safety measures described in Section 7.5.

7.2.2
Adhesive Mixing

The mixing of adhesives is conducive to the compounding of the components of a reactive adhesive in the prescribed ratio to trigger off the chemical reaction of curing.

7.2.2.1 Industrial Processing

In industrial adhesive processing, the mixing process of the prescribed portion occurs directly prior to the application in combined mixing, dosing and application devices.

Compared to manual mixing, this proceeding bears the following advantages:

- semi- or fully automatic processing,
- no exceeding of pot life,
- mixing ratio continuously adjustable,
- no adhesive losses, since only the quantity required is mixed,
- no danger of skin contact with the adhesive,
- no errors in the quantity ratio of the mixture,
- homogeneous, bubble-free adhesive mixture,
- exact dosing,
- very high repeat accuracy,
- automatic cleaning of the equipment.

7.2.2.2 Application in Workshops

For the processing in workshops, laboratories or in the private sector, automated facilities are usually too expensive. Here, manual mixing of the adhesives is indicated. To avoid possible mistakes, it is advisable to observe the following instructions:

- Mix the adhesive only after the adherends and their surfaces have been adequately prepared and the devices for the ensuing fixing of the adherends after the adhesive application are available.
  Choose suitable adhesive batches to avoid exceeding of pot life.
- In the case of adhesives with dyes added to one component, carry out the mixing process until a consistent shade of the batch is reached.

- It is advisable to use a stainless steel, glass or wood spatula and a clean working area (glass, aluminum foil) or a disposable plastic cup for the mixing of adhesives (preferably polythylene or polypropylene since plastics like polystyrene, polycarbonate or polyvinyl chloride may swell due to components of the adhesive).
  Tip: For frequent bonding processes with low adhesive consumption, even so-called "pill cups" (content approximately 20 ml) of polyethylene or polypropylene are suitable, which are available in drug stores.
- Do not mix too quickly, to avoid air bubble inclusions.
- When the components require weighing, it is recommendable to dose them first side by side on the mixing area, or if containers are used, oppositely to be able to remove excessive quantities, if any.
- During weighing, never put an spatula contaminated with one of the components into the container of the other component.
- Collect the uncured adhesive residues in closable containers and dispose of them as hazardous waste.
- Cured adhesive residues are not reusable. There is no use in trying to make it ready for use again by adding solvents.
- Cover the bonding place with paper or aluminum foil.
- For the cleaning of the stainless steel and glass spatula, paper tissues soaked in acetone are suitable.

7.2.2.3 Dynamic Mixers

There are different preconditions for the mixing of adhesives:

- Very different portions of the components, for example, only a few per cent of hardener in the resin component of methacrylate adhesives.
- Components with large viscosity differences.

In these cases, mixing is preferably carried out by means of a *stirrer*. Since here, the rotation of the stirrer is responsible for the mixing process, this kind of mixing device is also called a dynamic mixer (from Greek *dynamicos* = moving, effective) (Figure 7.9).

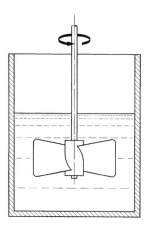

Figure 7.9 Dynamic mixer.

7.2.2.4 Static Mixers

In the case of adhesives without extreme mixing ratios and/or viscosity differences, mixing is carried out in *static mixers* (from Greek *statos* = standing, immovable). Inside these devices, a mixing helix is fixed in a (mixing) tube and offset by angles of 90° (Figure 7.10).

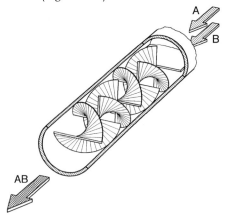

Figure 7.10 Static mixer.

The components to be mixed are separated into two branch currents at the leading edge of the first coil of the mixing helix. At each of the following mixing coil edges, these two branch currents are again separated. Depending on the number of mixing coils in the mixing tube, the repeated separation of the layers may result in a very intensive mixture, the desired consistency of which can be calculated in advance by the number of mixing coils. With 10 coils, 1024 layers are already achieved (Figure 7.11).

The mixed adhesive discharges at the front end of the mixing tube. To be able to process adhesives in this way, adhesive manufacturers offer the components A and B in two separate cartridges, from which activated feeding plungers press them into the mixing tube, where they are then mixed (Figure 7.12).

When finishing the bonding process, it has to be considered that the adhesive remaining in the mixing tube will cure there, and the mixing tube will not be usable again. For this reason, it is advisable to plan and prepare all bonding processes to the effect that they can be carried out in a single process step. One way to maintain

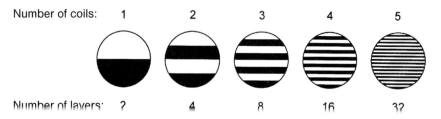

Figure 7.11 Layer formation in a static mixing tube.

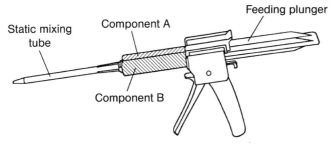

Figure 7.12 Manually operated "hand-pistol" for the processing of two-component adhesives.

the serviceability of the mixing tube filled with adhesive is keeping it in a deep freezer after finishing the work. As described in Section 3.1.4, the components' disposition to react with each other is limited at low temperatures so that the pot life in the mixing tube extends; however, to what extent depends on the adhesive and has to be determined by a test.

Static mixing tubes are particularly suitable for the processing of two-component epoxy resin adhesives, polyurethanes and methacrylate adhesives (A-B method, Section 4.3.3) with mixing ratios in equal shares adjusted by the manufacturer.

An advantage of mixing with a static mixing tube is the fact that no air is introduced into the adhesive.

For single applications in the repair sector, reactive adhesives in tubes or booklets separated by sealed seams are available. Here, the components are squeezed out in equal strand lengths and mixed by an spatula. At any rate, it is important not to exceed the pot life indicated on the package. After squeezing out the components from tubes don't change their closures by mistake, otherwise they will be "adhesively bonded" to the tubes.

The following information is generally applicable:

- Cold (at room temperature) curing adhesive systems have short pot lives (in the range of seconds, minutes, hours).
- Heat (at temperatures of approximately 60 to over 100 °C) curing adhesive systems have longer pot lives (hours, days, with cold storage even weeks, as the case may be).
- Pot life can be extended by cooling the adhesive batch.
- Pot life depends on the quantity of batch to be mixed (Section 3.1.1).

7.2.3
Adhesive Application

In many cases, there is no clear delimitation between application and mixing of the adhesives. In particular, in the case of adhesive systems with very short pot lives, mixing, dosing and application devices are often one single unit. Investing in such equipment is not only sensible for automation purposes, but also results in savings in adhesives since wrong batches or exceeded pot lives can be avoided.

7.2.3.1 Application Methods

The following application methods (Figure 7.13) are possible in the given order with increasing adhesive viscosity (with the adequate pressure, even adhesives with very high viscosities are processable with spray guns):

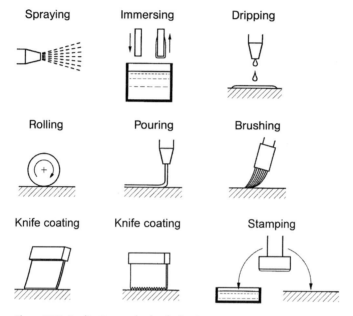

Figure 7.13 Application methods of adhesives.

The application of the adhesive is carried out in form of spots, lines, or surface areas. When determining the application method, the following criteria have to be considered, *inter alia*:

- kind of adhesive (one or two-component, pot life, mixing ratio of the components, viscosity, moisture sensibility (cyanoacrylate, polyurethane), possible filling material, required heat supply);
- adhesive quantity to be applied;
- desired degree of automation, application speed;
- spot, line or surface application;
- form of the bonding surface;
- accuracy of the dosing quantity.

7.2.3.2 Laminating

Laminating means the large area, continuous joining of flexible foils by means of bonding. It enables the production of composite material by combining different functional properties of the basic materials. The most common composites are made of paper, aluminum, polyethylene, polypropylene, cellophane and polyester, for example in the packaging industry. Laminating is also possible with thicker adherends, such as boards made of wood or plastic, cardboards, veneers, fabric and the like.

Laminating is, furthermore, important as a manufacturing process for composite materials by means of so-called laminating resins (unsaturated polyester, epoxies with substrates such as glass and carbon fibers) for example, in aeroplanes, vehicles and in shipbuilding. In these applications, however, bonded joints in the true sense of the meaning do not occur.

Independent of the application system to be selected, the following aspects should be considered:

- If possible, carry out the adhesive application immediately after surface pretreatment.
- See that the surfaces are equally wetted by the adhesive.
- The application of the adhesive to both adherends bears the advantage of equal wetting conditions. Quickly drying solvent-based adhesives should generally be applied to both adherends.
- Depending on their thermal conductivity, the adherends should be preheated (to the temperature of the melt) prior to the application of hot-melt adhesives.
- In the case of solvent-containing adhesives, a minimum drying time has to be provided for. This is particularly applicable if both adherends are impermeable to solvents.
- It is reasonable to remove exceeding adhesive leaking from the glueline edges before curing, since in a cured state, this would require forces that might result in the bonded joint being mechanically damaged. Another reason is the manipulation of test results in comparable tests, due to cured adhesive sticking at the glueline edges.
- To protect neighboring areas of the adherend surface from being wetted by the adhesive, these areas can be covered by adhesive tapes. However, in the case of hot curing up to approximately 110 °C, temperature-resistant adhesive tapes are required.

7.2.3.3 Amount Applied

The answer regarding the question of the adhesive amount applied depends mainly on the roughness of the adherends. Figure 7.14(a) shows two adherends with an assumed roughness of 50 µm (0.05 mm). An adhesive layer joining these adherends in the given way can merely fill the "valleys", while the surfaces contact at their peaks and "penetrate" the adhesive layer at these points. Therefore, an equally developed adhesive layer is not given. Only the application of a larger amount of adhesive, as shown in Figure 7.14(b), leads to an adhesive layer no longer impaired by roughness peaks and therefore able to transfer the respective forces.

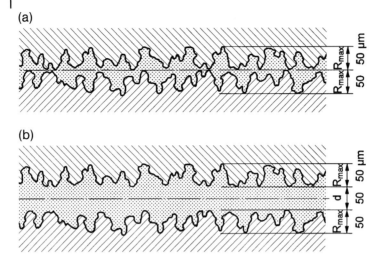

Figure 7.14 Relation between adhesive layer thickness and surface roughness.

A rule of thumb may be that the adhesive layer existing between the roughness peaks should correspond at least to the value of the maximum roughness, thus in the case of Figure 7.14 (b) 50 µm. Since depending on the treatment of the materials, usual roughnesses show values between 50 and 200 µm, adhesive layer thicknesses in this range are also common.

Moreover, it has to be noted that the adhesive layer thicknesses mentioned above provide for sufficient bond strength for most bonded joints. In special cases, for example, in car manufacturing, adhesive layer thicknesses in the range of millimeters are common for the bonding in place of windows or roofs (Section 10.3).

▶ **Supplementary Information:**

Sometimes processing instructions indicate the amount of the adhesive to be applied in "gram adhesive per m² of adherend surface". With an average specific weight of the adhesives of 1 g/cm³, the indication 100 g/m² corresponds to an adhesive layer thickness of 0.1 mm, respectively, 100 µm. This relation applies only to solvent-free adhesives. In the case of solvent-containing adhesives, the respective proportion of the solid content or polymer content has to be taken into account.

7.2.4
Fixing of Adherends

After the adhesive application, the adherends have to be fixed to prevent them from shifting against each other while curing. Shifting during the curing process leads to a distruction of the adhesive layer structure and thus to a distruction of its cohesive strength (Section 6.4). Usually, fixing takes place by means of pres-

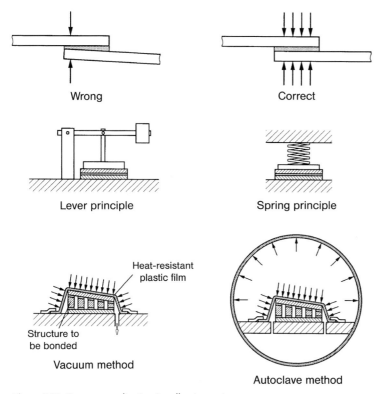

Figure 7.15 Pressure application in adhesive curing.

sure application to the adherends. Since equal adhesive layer thicknesses are an important precondition for the behavior of bonded joints, attention should be paid to equal surface stress when pressure is applied to the adherends (Figure 7.15).

For individual bonds, screw clamps and weights are suitable for pressure application, however, these technical aids are only reasonable for small numbers of items. For series bonds it is necessary to manufacture special fixing devices according to the geometry of the adherends.

The fixing of adherends is also possible in a simple way with adhesive tapes that, in case of hot curing, have to be heat resistant.

To be able to determine the required contact pressure for an intended adhesive layer thickness in pretests, the insertion of wires with diameters corresponding to those of the adhesive layer thicknesses in the area of the overlap ends has proven its worth.

7.2.5
Adhesive Curing

When discussing this topic, it is advisable to differentiate between two terms: *drying* and *curing*.

7.2.5.1 Drying, Evaporating

Drying or *evaporating* are discussed after the application of solvent-containing or aqueous adhesives/dispersions. After evaporation, respectively, penetration of the solvent or water into the porous adherends, accelerated by a heat supply if necessary, the adhesive layer polymers remain in the glueline. As described in Section 2.2.2, this is a physical process. A chemical reaction does not take place.

7.2.5.2 Curing

The term curing describes the transition of a reactive adhesive from a liquid or even paste-like state to the solid adhesive layer through chemical reaction. Here, compulsory temperatures and times have to be observed. The temperatures are measured and, if required, automatically recorded at the sample to be cured by means of sensors. Heat should be supplied constantly and equally, for example, in a circulating air oven. If heated too quickly, the reduced viscosity could cause the adhesive to leak out of the glueline before curing sets in. The curing time is the period of time in which the prescribed curing temperature prevails, without heating and cooling times.

In practice, the temperature–time curve schematically shown in Figure 7.16 does not only depend on the adhesive-related parameters, but also on the properties of the adherends, especially on their thermal conductivity. High thermal conductivity (e.g., metals) leads to shorter heating times than low thermal conductivity, as is to be found, for example, with plastics, glasses, wood. Even the dimensions of the adherends play a role.

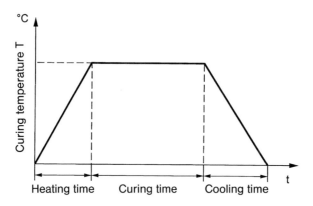

Figure 7.16 Temperature–time run of a curing process of reactive adhesives.

In conclusion, based on Sections 7.1 and 7.2 it has to be established that bonding, especially on an industrial scale, is a manufacturing process comprising the following process steps with regard to the adhesive:

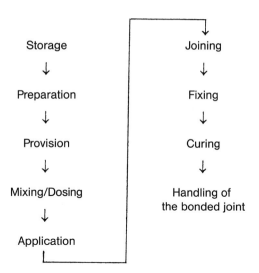

Each individual process step is of decisive importance for the total quality of the bonded product. In this context, it may be permitted to define an adhesive as a *process material*. This demand derives decisively from the fact that nondestructive testing for bonded joints is only possible to a limited extent and/or their realization involves a great metrological effort.

The maxim is:

> *Quality is not a result of testing, it must be an integral part of the manufacturing system!*

7.3
Repair Bonding

7.3.1
Metal Components

In many cases, bonding enables the repair of damaged workpieces or components of metal and nonmetal materials. The decisive advantages of repair bonding are the favorable ratio of repair costs to acquisition costs of new components, the shortening of downtimes, application even in environments with inflammable substances and thus no dismounting of the repair part.

Independent of the individual repair case, the practical work has to comply with the known rules for the bonded joint production:

- At first, it has to be ensured that the spot to be repaired is dry and free of contaminations from the defect component (if required, turn the component, remove the residues and dry the component).

- Mechanical removal of adhesive layers (sanding, rotating steel brushes, etc.) with ensuing degreasing has to be carried out as surface pretreatment.
- It is reasonable to provide a repair-bonding surface exceeding the actual damaged spot.
- If possible, further crack propagation should be limited by placing a bore.

Figure 7.17 schematically shows a repair bonding in the case of a crack in a thick-walled metal component.

Cold curing two-component reactive adhesives, for example, on an epoxy-resin basis are preferably used. Since the damages to be repaired are mainly cracks or surface defects with larger gap widths, the adhesive should show the respective gap bridgeability, which is achieved by adding filling materials. To avoid internal stresses, it is recommendable to choose filler materials similar to the adherends (e.g., steel powder, aluminum powder, bronze powder). In this manner, the thermal expansion coefficients of the joint and the component material can be adapted to a large extent. With regard to form and surface condition, the repaired area can be largely adjusted to the original component by mechanical processing methods (filing, sanding, etc.). The products available in the trade take account for these requirements.

In cases requiring crack sealing, one can proceed as shown in Figure 7.18.

After the respective surface treatment, a precut part made of characteristic material is glued over the damaged spot. To reinforce the adhesive layer, fiber-glass cloth can be included by laminating. Round components require the prior

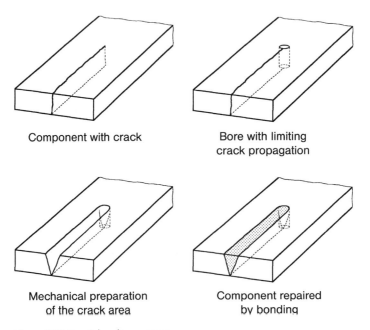

Figure 7.17 Repair bonding process.

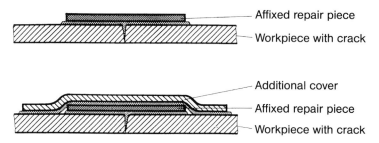

Figure 7.18 Crack repair by surface cover.

rounding of the part to be glued and high stiffness to eliminate peel stress in the bond overlap regions. In particularly critical stress cases (e.g., internal pressure in the container to be repaired) it is recommended to use another solidifying board (Figure 7.18, lower picture).

Repair bonding is of special importance in car manufacturing for the "elastic bonding" applied there (Section 10.3). In the case of damage, bonded windows or side parts have to be replaced. Since the adhesive layer thicknesses are in the millimeter range, repair is carried out by special dismounting tools, for example, by cutting wire or pneumatically or electrically operated vibration knives. The adhesive layer residues remaining after the damaged part has been cut out, represent a sufficient base surface for the application of a primer adjusted to the adhesive, so that complete removal is not required. The new adhesive can be directly applied.

7.3.2
Plastics

7.3.2.1 Rigid Materials

Here, in particular, reinforced plastics like GF-UP (molding material of glass-fiber reinforced, unsaturated polyester resin) and SMC (laminar semifinished parts of glass fabric, fillers and unsaturated polyester resin) are concerned, as used in car and ship building. For these applications, suitable repair sets on the basis of unsaturated polyester resins are available that enable repair according to the following procedure:

- Mechanical abrasion in the damaged region. Removal of overlapping fibers, if any, removal of sanding dust, premixing of the resin-hardener component according to the manufacturer instruction (pot life approximately 15–20 min). A glass fabric blank in the size of the damaged spot is put onto the prepared spot and impregnated with a resin–catalyst mix by means of a brush. The resulting laminate is freed from possible air-bubbles using a small anvil roller before final curing occurs (approximately 4–6 h). During the subsequent grinding, care should be taken that the glass fabric is not damaged. Gloves and protective goggles should be worn at any rate!

- Such repair work can be carried out considerably faster with light-curing glass-fabric-reinforced products. At first, a light-curing fiber paste from the tube is applied (to compensate possible unevenness), then a light-curing laminate (packed in a light-tight aluminum bag) is pressed into the surface prepared with fiber paste. Then curing takes place within approximately 10–15 min by means of a UV(A) lamp. Preparation and after-treatment occur in the way described before. The advantage of this system is the fact that component mixing does not take place and thus adherence to a prescribed pot life is not required.

7.3.2.2 PVC Films

Films of plastified polyvinyl chloride (PVC-flexible) are often used for the manufacturing of recreation-oriented products (boats, balls, rainwear, etc.). In the case of damage, repair work based on the principle of diffusion bonding (Section 9.2.5) is possible. Most repair systems available in the trade are adhesives consisting of the solvent tetrahydrofuran (THF) with portions of PVC powder and of pieces of PVC film in the respective colors. Repair is then carried out by roughening (sanding paper, fine wire brush) of the surfaces to be repaired (slightly larger than the piece of film to be glued), removal of residues, application of the adhesive to both surfaces, exhausting around the product (approximately 2–3 min) and strong compression.

Tip: To avoid tensions in the repair bonding, repair under inflated conditions is recommended.

7.3.2.3 Gummed Fiber Fabric

These materials, especially used for heavily stressed inflatable boats, cannot be repaired with the adhesive described for PVC in the previous paragraph, since the rubber coating will not be sufficiently swollen. In addition, the strength of the bonded joint is poorer than that of the coated fiber fabric, and thus the repair spot remains a constant "weak spot". Suitable repair adhesives for this application are

- two-component polyurethane adhesives (observe pot life) (Sections 4.2.1 and 4.2.5);
- contact adhesives (Section 5.3).

Repair occurs in the way described in Section 7.3.2.2.

> ■ *Most mistakes in bonding demonstrably occur because the production requirements for demanding bonded joints are not met, since technical, physical and chemical interactions are not known and thus cannot be taken into consideration. Quality defects of the processed adhesives as a reason for the failure of a bonded joint are hardly ever verified!*

7.4
Mistake Possibilities in Bonding and Remedial Actions

The following table helps to detect reasons for mistakes in bonding and thus to avoid their possible recurrence.

1. Uneven surface wetting by the adhesive

Possible causes	Remedial actions
1.1 Surface contaminations through fats, oil, solids (dusts)	• Carry out or repeat surface treatment
	• Check compressed air used in blasting for freedom from oil
	• Do not touch surface with your hands after surface treatment (cotton gloves)
	• Check solvent used in degreasing for fatfreeness (replace solvent, if required, vapor degreasing)
	• Check whether fat residues are able to be removed by the applied solvent (some fats are hardly soluble)
1.2 Moisture condensation on the surface due to temperature differences	• Climatization of the adherends
1.3 In the case of plastics, plasticizers diffusing on the surface	• Remove by means of surface treatment (mechanical)
1.4 Adhesive viscosity of solvent-based adhesives too high	• Readjust adhesive viscosity by means of suitable solvents or diluting agents
1.5 Adhesive viscosity of reactive adhesives too high due to exceeded pot life	• Hold new adhesive mix available. Dilution of unusable adhesive mixes with solvents will not be at all successful! Thus let adhesive mix cure completely and dispose of it properly
1.6 Inhomogeneous adhesive mixture (in the case of filler-containing adhesives)	• Mix adhesive again
1.7 Residues of protective paper or film	• Carry out or repeat surface treatment
1.8 Melt viscosity of hot-melt adhesives too high	• Increase melt temperature. If required, preheat adherends (in case of metals)

2. Insufficient adhesive properties of the adherends and occurrence of adhesive fractures

Possible causes	Remedial actions
2.1 See 1.1, 1.2, 1.3, 1.7	
2.2 Exceeded pot life in case of reactive adhesives	• Use new adhesive mix
2.3 Insufficient adhesion of the layers already existing on the adherends (lacquers, corrosion protection layers, metal layers)	• Remove layers mechanically, degrease and primer the surfaces, if required
2.4 Adhesive layer thicknesses too small due to the removal of liquid adhesive in case of porous adherends	• If required, apply adhesive a second time
	• Use adhesive with higher viscosity

3. Insufficient cohesion strength of the adhesive layer

Possible causes	Remedial actions
3.1 See 1.6, 2.2, 2.4	
3.2 Incomplete or uneven curing of the adhesive layer	• Check for possible deviations from the prescribed mixing ratio of the components
	• Check mixing and dosing equipment
	• Check time and temperature control in the case of hot-curing adhesives
	• See to homogeneous mixing in case of adhesives with added fillers
	• If required, extend curing time or choose higher curing temperature
3.3 In the case of fast-setting adhesives and large adherends, possibility of curing starting prior to fixation of the adherends	• Choose adhesives with longer open assembly times
	• Reduce time cycle
3.4 Insufficient moisture contents of the air when processing cyanoacrylate adhesives and one-component polyurethane adhesives	• Climatization of the adhesive processing rooms
	• In the case of cyanoacrylate adhesives, provide smaller adhesive layer thicknesses (approximately 0.1 mm)
3.5 Adhesive layer thicknesses too small or uneven	• See to flatness of the adherend surfaces, if required, remove flash at the edges of the adherends
	• Constant application of contact pressure
3.6 Entrapped air or solvent in the adhesive layer	• Mixing under vacuum, if required, reduce stirring speed

7.5
Safety Measures in Adhesive Processing

Similar to other manufacturing processes, bonding requires measures for the protection of man, company and environment. In contrast to welding and soldering, in bonding organic products are almost exclusively utilized, which can be subsumed under different hazard categories for the purpose of health and fire protection. Due to the variety of existing formulation components and processing methods, there is no possibility to assign product-related characteristics regarding processing regulations to be observed to individual adhesives.

In this context, the reader is referred to the Safety Data Sheet For Hazardous Substances and Preparations (TRGS 220 according to 91/155 EEC as well as its amendment 2001/58 EC) provided by the European Community. Its aim is to communicate the physical, safety-related, toxicological and ecological data of the individual products essential in the handling of chemical substances and preparations, as well as to give recommendations on safe storage, handling and transport. Although it is not intended for the private end user, it still provides the possibility to obtain supplementary information of the manufacturer for the industrial application of adhesives. Furthermore, in-house prepared data sheets provide the user with important information on the part of the adhesive manufacturer.

7.5.1
Workplace Prerequisites for Adhesive Processing

1. Sufficient air ventilation or exhaustion. In the case of exhaustion systems, it has to be considered that solvent vapors are heavier than air, thus exhaust at ground level, too.
2. Provide fire extinguishers (powder extinguishers). Do not try to extinguish with water, since solvents "float" on the water surface, thus further extending the fire.
3. Since the extent of a possible fire is determined by the range of combustible material, store adhesives and solvents only in quantities actually required at the workplace.
4. Provide well-marked and closable waste containers for solvents, adhesive residues, acids, bases and cleaning cloths. Noncured adhesive residues are regarded as hazardous waste!
5. Provide the following articles at each workplace or at a central place in the workroom:
   – industrial overall (cotton),
   – protective goggles,
   – expendable towels,
   – respirator mask,
   – skin creme,
   – eye wash,
   – body shower,

- absorptive material (diatomaceous earth, mica mineral, sand, if required) for spilled or leaked liquid products.
6. Reference to telephone numbers of
   - medical practitioners,
   - fire brigade.
7. Marking of hazardous substances by standardized danger symbols (usually on the packaging provided by the respective manufacturer). Important symbols are shown in Figure 7.19.

Highly inflammable   Toxic   Noxious   Caustic

Figure 7.19 Danger symbols (examples).

7.5.2
Rules of Conduct in Adhesive Processing

General Notes: It cannot be excluded that employees involved get into contact with chemical substances. This may happen by swallowing (orally), skin contact (dermally) and inhaling. While working consciously, the first two cases may be avoided, however, inhaling over a longer period cannot be avoided. Regarding probable risks, the following ranking of the importance of the respective detrimental effects results:

Inhaling more detrimental than *skin contact*, more detrimental than *swallowing*.

So-called MAK values (maximum workplace concentrations) have been defined for individual chemical substances as a preventive measure against health problems due to inhaling, which must not be exceeded at the workplace. After these preliminary remarks, the prohibition of eating and smoking at the workplace is a logical consequence. Furthermore, the following aspects have to be considered:

1. Wear protective clothing.
2. Do not fill substances in unlabeled containers, in particular not in containers intended for food (beer and water bottles).
3. Do not dispose of chemicals and solvents in drains.
4. Do not dilute acids and bases by adding water to counter the strong heating, but *vice versa*, pour the acids and bases slowly into the water while stirring carefully.

5. Slip off your clothes immediately, if chemicals are splashed, wash up possibly affected areas of skin using plenty of water, apply skin-protective lotion.
6. In the case of a chemical burn of the eye, keep it wide open with both hands and rinse it under running water or use eye wash. Afterwards, see a doctor immediately.
7. Keep the workplace clean.
8. When disposing of adhesives please note that:
   – Liquid or paste-like adhesive residues that are not cured or have exceeded the storage time required for a perfect processing, are generally regarded as hazardous waste. The same applies to packaging with respective adhesive residues.
   – Cured adhesives, for example, after exceeding the pot life, can be disposed of together with domestic waste.
   – Residues of solvent-based adhesives are to be disposed of as hazardous waste in well-closed containers provided with adequate labelling.

For further detailed information regarding this topic, refer to "Information series of the German Chemical Industry Fund, No. 27 Bonding/Adhesives" (www.vci.de/fonds).

7.6
Quality Assurance

As already mentioned in Section 7.2, manufacturing processes in general and, due to the only limited availability of nondestructive testing methods, bonding in particular, require process-accompanying quality assurance. A comparison with welding and soldering will certainly back this statement. In these two materially joined bonding processes, the quality properties of the additional material (composition of the alloy, metallurgical structure, etc.) are predefined by the manufacturer and are reflected in the finished bonded joint. In bonding, the adhesive layer develops only under the responsibility of the user and can be influenced in many ways by the given manufacturing conditions.

Certainly, the following summary will mainly concern industrial production, however, it can also be a useful help for the nonindustrial user, according to the title of this book "How to achieve flawless results".

Planning

- Training schemes for employees.
- Integration of the adhesive manufacturing into the design phase.
- Establishing of in-house regulations or standards.
- Selection of adhesives (Chapter 8).

Adhesives

- Verification of supplier information on the label to avoid confusion.
- Control of viscosity, density, solid content and color, if required, for additional adhesive identification. In the case of one-component reactive adhesives, the viscosity test allows the verification of a probable exceedance of pot life (gelification).
- Control of storage time and temperature regarding possible exceedance of pot life.
- Realization of test bondings and their verification, if required.

Adherend Material

- Examination of the surface condition (cleanliness, freedom from fat), verification of wettability (water-droplet test, Section 7.1.1.4).
- Roughness test.
- Dimensions, tolerances.

7.7
Adhesive-Bonding Training

As already mentioned at the end of Section 7.2.5.2, the successful application of bonding as a manufacturing process requires careful planning regarding staff and technical preconditions. Thus, the demand for quality management for the purpose of DIN EN ISO 9001: 2000, nowadays indispensible for industrial manufacturing, also includes bonding as a joining process.

Based on the year-long experience in welding training, a comparable system has been developed for the manufacturing system of "bonding", lead-managed by the *German Welding Society (DVS)*, Düsseldorf, and supported by industry and science. The respective training programs are defined in guidelines and regulations. This training system has been introduced Europe-wide. The qualification measures comprise several training stages. The most important guidelines and regulations are mentioned below (EWF = European Welding Federation):

Regulation	DVS-EWF 3301:	Training and Examination – Adhesive Specialist
Guideline	DVS-EWS 3302:	Adhesive Specialist – Basic Modul
Guideline	DVS-EWF 3303:	Adhesive Specialist – Advanced Modul Metalbonding
Guideline	DVS-EWF 3304:	Adhesive Specialist – Advanced Modul Plastic bonding
Regulation	DVS-EWF 3305:	Adhesive Bonder – Training course and Examination
Regulation	DVS-EWF 3309:	Adhesive Engineer
Regulation	DVS-EWF 3310:	Quality Management Adhesive Bonding

In addition:

Document	EWF 515-01:	Europoean Adhesive Bonder
Document	EWF 516-01:	European Adhesive Specialist
Document	EWF 517-01:	Europoean Adhesive Engineer

Complementary literature to Chapter 7:
[A1, B5, C1, F1, G1, M3, S1, W2].

8
Adhesive Selection

8.1
Preliminary Notes

The most frequently asked question in the field of adhesive technology is the question regarding *the* suitable adhesive for *the* bonding problem to be resolved. The questioner is then often disappointed, because a clear-cut answer is not possible. The uncertainty is enhanced by the almost infinite range of adhesives, but also by the "promises" frequently found on the packing saying that the application possibilities of the individual products for "bonding and joining of all kinds of materials" are almost unlimited. Combined with incomprehensible chemical terms – and maybe bad experience – it is no surprise that some user or other does not really trust in bonding.

The study of the previous chapters may hopefully convince critics that this modern joining process is based on solid production-related, chemical and physical principles that, if carefully observed, guarantee high quality levels. It has to be mentioned, however, that the preconditions for the application of bonding are quite different. There is, on the one hand,

- the industrially applied "manufacturing system of bonding", and
- "bonding", carried out in handicraft businesses and in the field of do-it-yourself and households.

In the following, we therefore attempt to provide both fields of application with suitable information, since successful and precise bonding is the aim of all users. In addition, adhesive manufacturers have a comprehensive range of information material on adhesive selection and adhesive processing for the products they offer.

In order to systemize the information important for the selection of the adhesive, the following preliminary notes are helpful:

1. The information is limited to the most important materials, metals, plastics, (thermoplastics, thermoset materials, foams), ceramics, glass and their possible combinations. For papers, cardboards, wood, rubber polymers, usually physically setting systems (solvent-based, dispersion, hot-melt adhesives) are utilized. In these cases, the adhesive selection with regard to the manufacturing conditions and demands is of less problematic nature.
2. The different possibilities of surface pretreatment are left unconsidered in the systematics of adhesive selection. Except for very special conditions regarding climate and humidity in case of long-term effects, which require expensive chemical and electrochemical treatment, it is assumed that the process combination:

    Degreasing – Mechanical Pretreatment – Degreasing

according to Sections 7.1.1 and 7.1.2 suffices for most applications.

8.2
Determining Factors for the Selection of Adhesives

Figure 8.1 shows the determining factors to be generally observed in the selection of adhesives.

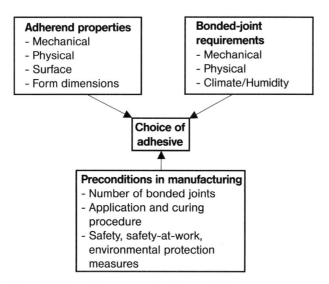

Figure 8.1 Factors determining the selection of adhesives.

8.2.1
Adherend Properties

	Examples
• Material (kind and composition)	
– hard, brittle, nondeformable	ceramic, glass
– elastic, plastically deformable	metals, thermoset materials, thermoplastics
– rubber-like flexible	rubbers
• Temperature stability	
– up to 100 °C	metals, ceramic, glass, thermoset materials, thermoplastics
– 100–200 °C	metals, ceramic, glass, thermoset materials
– over 200 °C	metals, ceramic, glass
• Insolubility in organic media – nonswellable	metals, ceramic, thermoset materials
• Solubility in organic media – swellable	thermoplastics, rubbers
• Thermal conductivity	
– high	metals
– low	glass, ceramic, thermoset materials, thermoplastics rubbers
• Surface	
– smooth	metals, plastics, ceramic, glass depending on manufacturing and surface pretreatment
– porous	plastic foams, cellulose materials, textiles
• Surface coating	see endnote[a]
• Adherend surface	in mm^2 range in cm^2 range in m^2 range
• Gluelines	
– plane	lap joint, butt joint
– round	tube, socket, shaft, collar joints
• Adherend combinations	metal/glass, metal/plastics

[a] Since, in the case of surface layers (metals, plastics, lacquers, paints) a firm joint with the basic material is not always guaranteed, it is advisable to remove them mechanically in the area of the adhesive layer and then to apply the adhesive directly to the basic material, if required use a primer.

8.2.2
Demands on Bonded Joints

- Not stressed by forces, fixation joints
- Mechanical stress by tension, tensile shear, pressure, torsion
- Exposure to moisture, climate.

Note: In the case of metal bonded joints, the resistance to moisture and climate can decisively be influenced by the kind of surface pretreatment apart from the adhesive selection, especially in the areas adjacent to the adhesive surface (e.g., primer, sealing of gluelines).

- Stress
  - by low temperatures −30 to 0 °C
  - by normal temperatures 0–60 °C
  - by elevated temperatures 60–120 °C
  - by high temperatures, above 120 °C

8.2.3
Preconditions in Manufacturing

- Quantities
  - Individual bonded joints
  - Small quantities (e.g., for test purposes)
  - Serial production
    - quantity per time unit
    - mechanized
    - automated
- Method of application (spraying, immersion, dripping, rolling, pouring, brushing, coating, knife coating, stamping)
- Application form (spot, line, surface)
- Mixing
- Mixing/Dosing
- Melting
- Preheating of the adherend
- Adhesive layer thickness
- Adhesive curing (cold, warm curing, radiation)
- Adherend fixation (short-/long-term)
- Successive manufacturing steps (temperature stress due to paint drying, mechanical agitation)
- Solvent disposal
- Exhaustion
- Increased fire protection
- Safety at work

8.2.4
Processing Parameters of Adhesives

- Cold, heat and hot curing
- One-component, chemically reacting
- One-component, physically setting
- Two-components, chemically reacting
- Pot life
  - below 5 min
  - 5–60 min
  - in the range of hours
- Open assembly time
  - short (range of minutes),
  - long (up to 60 min)
- Solvent-free
- Solvent-containing
- Curing time
  - below 5 min
  - 5–60 min
  - in the range of hours
- Viscosity
  - low 10–200 mPa s
  - medium 200–2000 mPa s
  - high 2000–20 000 mPa s
  - paste-like 20 000–over 100 000 mPa s (m = milli = 10^{-3})
- Gap bridgeability
- Adhesive layers hard, brittle, little deformable
  - elastic/plastically deformable
  - flexible
- Adhesive layers temperature resistant up to
  - 100 °C
  - over 100 °C
- Adhesive layers with fillers for certain demands (e.g., electrically/heat conductive)

8.2.5
Property-Related Parameters of Adhesives and Adhesive Layers

In this section, the most important properties of adhesives and the resulting adhesive layers, which have to be taken into consideration, are again summarized in a systematic scheme. For supplementary information, the reader is referred to the respective sections.

8.2.5.1 One-Component Reactive Adhesives

Epoxides:

- Heat/hot curing.
- For continuous stress of flexible materials (bending, rolling, suitable to a limited extent only).

Polyurethanes:

- Curing by moisture from the air and/or of adherend.
- For moisture-impermeable adherends only suitable if water addition (spraying, booster) is possible or adhesive application occurs in wave or line form to enable air access.
- Curing time depending on existing moisture, no "accelerated cure".
- Consider the possibility of CO_2-formation in case of thicker adhesive layers and high viscosities.
- Consider skin formation time (Section 4.2.3).
- Alternative: two-component PUR adhesives.

Silicone RTV-1:

- Predominantly applied as sealing compound.
- Curing by moisture from the air.
- Curing time depending on the layer thickness in the range of hours or days.
- In the case of commercially available formulations, separation of acetic acid by curing reaction, typical "vinegar odor".

Cyanoacrylates:

- Curing by moisture on the adherend surfaces.
- Very short open assembly time, fix adherends immediately after adhesive application.
- Adherend displacement after fixation possible only to a limited extent.
- Particularly suitable for small (mm^2, cm^2) adherend surfaces. Less suitable for larger (> approximately DIN A6) adherend surfaces due to the short open assembly time.
- For bonding of porous adherends use adhesives with gel-like consistency; in the case of the viscosity being too low, "penetration" into the surface area is possible.
- Apply only thin adhesive layers, since the moisture on the surface required for curing is insufficient for a complete cure (particularly at low atmospheric moisture).
- The relative air moisture should range between 40 and 70%.
- Wear protective goggles at all times and avoid skin contact.

Anaerobically curing adhesives:

- Cure by removal of atmospheric oxygen with metal contact at the same time, thus mainly suitable for bonded metal joints (shaft–hub, screws, bolts, etc.).
- For bonded joints of plastics special primers are required.

Radiation curing adhesives:

- Consider adjustment of the wavelengths of the emitter (emission) and adhesive (adsorption).
- Provide design suitable for bonding (Figure 9.3).

8.2.5.2 Two-Component Reactive Adhesives

Epoxides
Polyurethanes
Silicon RTV-2

- Consider mixing ratios of the components and, if required, curing temperature and time according to the manufacturers' instructions.
- Depending on the manufacturing conditions, mixing and dosing devices required, if necessary cartridge processing (Figure 7.12).
- In the case of cold-curing (room-temperature) adhesives, heat supply may reduce curing time.

Methacrylates:

Four different processing possibilities:

- Hardener additive (powder) to resin component.
- Apply hardener dissolved in solvent to one adherend, the resin component to the other adherend.
- Mixing of resin and hardener component.
- Separate application of the components to the respective adherends.

All four systems cure at room temperature.

8.2.5.3 Physically Setting Adhesives

Solvent-based adhesives:

- Consider open assembly time, minimum drying time, maximum drying time (Figure 5.2).
- Not recommended for solvent-impermeable adherends.
- Contact pressure required.
- Consider combustibility of solvents.
- Adjust viscosity to surface structure, for example, pores, otherwise risk of "penetration", if required, repeated adhesive application.
- Consider dissolving properties when bonding plastic.

Contact adhesives:

- High contact pressure required, pressure more important than contact time.
- Particularly suitable for flexible materials in the case of bending and rolling stress.
- Long open assembly time, thus advantageous for large area bonded joints.
- After fixing the adherends, no displacement possible.
- Also available as two-component systems.
- Limited heat resistance up to approximately 80 °C, since there are no crosslinked polymer structures in one-components systems.
- Application possibilities for adherends with smooth and porous surfaces.

Dispersion adhesives:

- Mainly applied to wood materials.
- Setting by evaporation of water or penetration into the adherends.
- Not suitable for materials with smooth and impermeable surfaces.
- Consider frost sensitivity, no longer usable after thawing.
- Setting time increases according to the moisture content of the adherends, since increasing humidity of the adherend (e.g., wood) delays the release of water from the dispersion.

Hot-melt adhesives:

- Very short open assembly time, fix adherends immediately.
- For the extension of the open assembly time of thermally well conductive adherends (metals), preheating up to the approximately melt temperature is required.
- Beware of burns; melting temperatures are over 120 °C.

Plastisols:

- Setting by sol-gel conversion under heat.
- Deformable adhesive layers.

Adhesive tapes:

- In many cases, alternative application possibility to liquid adhesives.
- Application possibility for adherend fixation when processing liquid adhesives.
- Advantages: clean processing, systems with high strength characteristics commercially available.
- Immediate handling strength, no displacement possible after fixation of adherends.

8.3
Selection Criteria

Based on the factors influencing the adhesive selection described in Sections 8.2.1–8.2.5 the last step is finding the "adequate" adhesives. Here, it is deliberately abstained from showing the usual selection lists for adhesives depending on the materials to be bonded, since such depiction does not provide sufficient space for additional and explaining information. In addition to the information given in Chapter 9, "Adhesive Properties of Important Materials", the author nevertheless acts on the assumption of a practice-related presentation of this important issue. The following criteria take the most important adhesive parameters into account and shall serve the user as a targeted orientation guide for practical implementation. Preconditions are that

- with regard to their kind of material and geometrical form, the adherends to be bonded are predetermined by the component to be manufactured, and that they are prepared according to the standards of a design suitable for bonding (Chapter 11) and
- the surfaces are in a "ready to be bonded" condition, due to the surface treatment already carried out (Section 7.1.2).

> ■ *It generally applies that the quality of an adhesive depends on the further processing of the adhesive itself and the adherends to be bonded. Failures are mostly attributable to improper processing conditions and not to the adhesive itself.*

1. *Strength of the bonded joint:* In the true sense of the word, the "load-bearing behavior", that is, the property of transmitting load. Influenced by
   - degree of crosslinkage of the adhesive layer (depending on the curing temperature) and based on this, its deformability;
   - constructive design (Chapter 11);
   - see Section 10.2.1 for definition of strength.

2. *Deformability of the adhesive layer:* Depending on the degree of crosslinkage, a guideline may be
   - high degree of crosslinkage: hard, partly brittle, little deformable adhesive layers; application in the case of adhesives strengths in the range of approximately 20–30 MPa, respectively, N/mm^2, epoxy resins, phenolic resins. See Section 10.1 for explanation of the dimension MPa;
   - medium degree of crosslinkage: in case of mechanical stress, reversibly partly even irreversibly deformable (creeping, Section 3.3.5). Bonding strength values in the range of approximately 10–20 MPa, methacrylates, polyurethanes, cyanoacrylate, anaerobic adhesives, hot-melt adhesives;
   - low, wide-meshed crosslinkage: adhesive layers with great reversible elasticity with bonding strengths of up to approximately 10 MPa, polyurethanes with weak crosslinkage, rubber and silicone polymers, acrylates (e.g., contact adhesives, pressure-sensitive adhesives).

▶ **Supplementary Information:**

The strength values mentioned above depend on the respective construction conditions as well as on the stress duration. An example: A plastic hook fixed to a tile by means of a pressure-sensitive adhesive and exposed to stress can come off in the course of time due to a failure of the adhesive layer (creeping). In the case of dynamic stress, the hook can break inside, the bonded joint remains unaffected. In this case, the adherend is the "weaker link in the strength chain".

3. *Exposure to moisture and climatic conditions:* The resistance of bonded joints against *exposure to moisture and climatic conditions* plays a role mainly in metal bonded joints, because of possible corrosion creep (Figure 7.8). Surface treatment even outside the glueline or – in extreme cases – sealing of the glueline edges helps to avoid this kind of failure. Highly crosslinked polyaddition and polycondensation adhesive layers show a lower moisture absorption than thermoplastic polymerization adhesive layers.

4. *Thermal stress:* Regarding the *thermal resistance* of the bonded joints, thermosetting adhesives are to be preferred to thermoplastically curing adhesives. The following information is useful for the determination of thermal stress:

Low temperatures	up to –30 °C	Polyurethane adhesives, silicon adhesives (if elastic adhesive layer properties are required)
Normal temperatures	0–60 °C	Practically all reactions and physically setting adhesives
Elevated and high temperatures	60 °C up to over 120 °C	Heat-curing reactive adhesives (epoxides, phenol resins)

5. *Adhesion:* Regarding the development of adhesive forces, the adhesives described differ only marginally. The most important parameter is basically the respective condition of the adherend surface to be bonded.

6. *Bonding of metal materials:* In the case of bonding with hot-melt adhesives, metal materials require preheating to the melt temperature to develop sufficient adhesive bonded joints, due to the high thermal conductivity.

7. *Bonding of thermoplastics:* For the bonding of thermoplastics (ABS, PVC, PC, PS, PE, PP) the following specific characteristics apply in comparison to metals:
   – The lower thermal resistance limits the application of heat or hot-melt adhesives with longer (approximately 12–15 min) curing times due to possible adherend deformation.
   – For the same reason, when bonding with hot-melt adhesives it is recommended to adjust their melting temperature in processing to the heat resistance of the respective plastic material.

8. *Bonding of thermoset plastic materials:* Due to their insolubility, thermoset plastic materials (e.g., items made of Bakelite, expoxy resin, boards with melamine urea coatings) are not bondable by surface dissolving with solvent-containing adhesives.

9. *Bonding of adherend combinations:* In the case of adherend combinations the following rules have to be observed:
   – Bonding of rigid, weakly deformable materials with plastic, flexible materials, for example, metal–rubber: In this case, choose adhesives that develop flexible, deformable adhesive layers, for example, low crosslinked polyurethanes, contact adhesives.
   – Materials with different thermal stressability, for example, metal–plastic, glass–plastic: Application of cold-curing adhesives or those with a curing temperature corresponding to that of the more temperature-sensitive adherend.
   – Apply adhesives with elastic adhesive layers for materials with different thermal expansion coefficients.

10. *Bonding of materials with solvent-impermeable surfaces:* Solvent-containing adhesives are less suitable for materials with solvent-impermeable surfaces (metals, glasses, thermoset materials), since the solvents still present prior to the fixing of the adherends cannot or can only slowly escape over the glueline edges, thus a strong adhesive layer is not possible.

11. *Bonding of porous materials:* For porous adherends, higher adhesive viscosities have to be chosen – depending on the pore diameter – to prevent the adhesive from penetrating (disappearing in the pores); if required, repeat adhesive application after a short drying time.

12. *Adhesive viscosity:* The viscosity of an adhesive has also to be adjusted to the development of the surface. Rough surfaces require generally lower viscosities than smooth surfaces to ensure even wetting. For the production of "thicker" adhesive layers in the range of millimeters, only filler-containing adhesives with very high viscosities are suitable.

13. *Pot life of adhesives:*
   – Adhesives with shorter pot life are only processable in serial bonding by means of automatic mixing and dosing systems and fixing of the adherends at the same time.
   – High expenditure for mixing and dosing devices are mostly compensated by lower adhesive costs (no losses due to excess pot life) and a higher quality standard of the bonded joints.

14. *Cure time, respectively, setting speed:*
   – Adhesives with short cure time or setting speed (cyanoacrylate, reactive adhesives with short pot lives, hot-melt adhesives) are only suitable to a limited extent for more extensive bonded joints (dm^2/m^2), since the application time may exceed the cure time. The bondable surface depends on the

"open" assembly time. It is generally applicable: Small adherend surfaces (in the range of mm^2–cm^2) are bondable with fast-curing adhesives. Extensive adherend surfaces (in the range of m^2) require adhesives with long open assembly times, respectively, pot lives.
- Efficient serial bondings with high production speed are only possible with fast-curing, respectively, setting adhesives. Slowly curing adhesives require costly devices for the fixing of the adherends.

15. *Atmospheric moisture:* Cyanoacrylate and one-component polyurethane, silicone adhesives require sufficient moisture in the ambient air (approximately 30–70% relative air moisture) for curing.

16. *Radiation curing:* For the application of radiation curing, at least one adherend has to be suitable for UV radiation. It is, furthermore, applicable that the diaphaneity of a material cannot be equated with its transparency for UV-rays. The radiation energy actually available for radiation curing in the glueline can be determined by means of a UV measuring instrument.

17. *Fire protection:* In contrast to solvent-free reactive or hot-melt adhesives, the processing of solvent-containing adhesives requires expensive measures for fire and explosion protection as well as for the disposal of the solvents.

18. *Adhesive tapes:* Apart from the application of the liquid adhesive systems described, it should not be forgotten that many bonding problems can be solved by double-sided adhesive tapes. The development of these systems, especially on the basis of adhesive foam structures or thermal post-curing, has made great advances in the past with regard to strength and possible stresses.

In conclusion, regarding the *technological properties* of the adhesives to be chosen, it has to be differentiated:

- According to the bonding properties indispensable for the application purpose. They depend on the respective polymer structure and represent the criterion for *binder, respectively, basic material*.
- According to the bonding properties indispensable for the handling in production. In general, they have to be coordinated with the adhesive manufacturer and represent the *processing criterion*.

With regard to the *economic requirements* of adhesive selection, it is a general rule to apply the adhesive, which offers the optimum solution of the bonding task and fulfills the quality requirements in a most efficient way.

Complementary literature Chapter 8:
[D1, P3, S1, S3].

9
Adhesive Properties of Important Materials

The basics and criteria for the selection of adhesives discussed in Chapter 8 refer, to a large extent, to process applications on an industrial scale. Experience shows that when choosing adhesives, users in the nonindustrial sector orientate themselves more on the bondable materials and their adhesive properties. For this reason, the following descriptions of the materials include information on the recommended or even not recommended adhesives with the respective justifications. In addition, it has to be mentioned that a large part of the explanations in Section 8.3 are of universal validity.

9.1
Metals

9.1.1
Fundamentals

An essential part of all bonding processes is carried out with metal materials – independent of the branch of industry. Thus, it is also necessary to know their adhesive properties in contrast to nonmetal materials. In general, experiences gained from *one* metal material, considering the explanations in the Chapters 3, 4, 8 and 11, can also be transferred to new tasks with *other* metals.

The bonding properties of metal materials are mainly determined by the following characteristics:

9.1.1.1 Strength
Most metals show a low deformability in comparison to nonmetal materials. For bonded joints this property means that adhesive layers, when exposed to mechanical stress (tension, shear, pressure, bending, torsion), are subjected to deformation stress only to the same extent.

Applied Adhesive Bonding: A Practical Guide for Flawless Results. Gerd Habenicht
Copyright © 2009 WILEY-VCH Verlag GmbH & Co. KGaA, Weinheim
ISBN: 978-3-527-32014-1

9.1.1.2 Impermeability Towards Solvents

This is a property metals share with materials such as glass, special plastics (above all thermoset materials) and partially even ceramics, which limits the application range of adhesives. According to the explanations given in Section 5.2, the maximum drying time to be observed after adhesive application and prior to the fixing of the adherends is a decisive parameter for the production of strong, load-transferring adhesive layers. If this time, which also depends on the adhesive amount applied, is not exactly observed, solvent residues can be entrapped thus reducing the strength of the bonded joint. Due to the impermeability of the adherends towards solvents, such components cannot escape afterwards, as is the case with porous materials. The same limitation applies to dispersion adhesives.

9.1.1.3 Insolubility in Solvents

Metal surfaces show a so-called "inert" behavior towards adhesives and solvents, that is, neither dissolving nor diffusion processes take place. This property requires a delimitation towards adhesives used for bonding of thermoplastics according to the diffusion bonding principle (Section 9.2.5).

9.1.1.4 Thermal Conductivity

The thermal conductivity of adherends influences the temperature conditions in the glueline during adhesive curing. It plays a special role in the application of hot-melt adhesives on metals due to the quick solidification of the melt in the boundary layer zone and the possible impairment of the adhesion development. The thermal conductivity λ is indicated in the dimension W/cm K (Watt per centimeter Kelvin). Values of certain materials:

Aluminum 2.3	Iron 0.75	Copper 3.8
Brass 1.1	Silver 4.2	Stainless steels 0.2–0.5
Glasses 0.01	Plastics 0.002–0.004	

9.1.1.5 Temperature Resistance

The high resistance of metal materials to thermal stress allows the application of reactive adhesives, which cure at elevated temperatures and show high bond strengths (up to 40 MPa).

9.1.2 Surface Pretreatment

For the production of bonded joints with metal materials, appropriate surface pretreatment is of priority. In the technical literature, various formulations of pickling solutions are to be found, their application, however, is limited for reasons of occupational safety and due to the disposal problem. Therefore, we refrain from describing them here.

In contrast to this, mechanical surface pretreatment methods, as described in Section 7.1, are universally applicable. With the process steps

> Degreasing – Sandblasting, respectively; Grinding, respectively, Brushing – Degreasing

being considered, and, if required enhanced by the sealing of the glueline edges to avoid corrosion creep, durable bonded joints are possible for almost all applications.

9.1.3 Bondability of Important Metals

Below, the most important bonding properties of specific metal materials are described with regard to their practice-related application.

9.1.3.1 Aluminum and Al-Alloys
- Base metal, that is, if stored, layers of different chemical compositions (oxides, hydroxide, oxydhydrates, carbonates) cover the surfaces; their adhesion to the base metal does not then guarantee sufficient strength for a bonded joint. Mechanical surface pretreatment is also required.
- Strong adhering reaction layers (Figure 7.7) are obtainable only with chemical or electrochemical treatment methods.
- High thermal conductivity.
- Most important alloys: Al Mg 3, Al Mg 5, Al Cu Mg 2 (aerospace industry).

9.1.3.2 Noble Metals
- The noble metals silver, gold, platinum are characterized by similar behavior in bonding. Their noble character enables processing without chemical surface pretreatments.
- Mechanical surface pretreatment, followed by very careful degreasing. Carry out bonding immediately, since especially silver surfaces can change due to silver sulfide formation (dark staining).

9.1.3.3 Stainless Steels
- The specific problem in bonding of stainless steel is its passiveness, that is, heavily reduced capacity of reaction towards affecting media. This property explains their application in the case of corrosion stress. Characteristic of bonding is the limited development of intermolecular bonds (Section 6.1).
- Mechanical surface pretreatment, advantageous with SACO-method (Section 7.1.2.1).

9.1.3.4 Copper
- Very high thermal conductivity.
- Easily deformable, especially Cu sheets, thus deformation characteristics of adhesive layers important for adhesive selection.

- Depending on the metallurgical condition of the copper regarding existing alloy elements, such as, for example, zinc (brass), nickel (coinage metals) the application of heat-curing adhesives can lead to a recrystallization and thus to reduced strength.
- Mechanical surface pretreatment.

9.1.3.5 Brass

Here, the characteristics mentioned in connection with copper are basically applicable.

9.1.3.6 Steels, General Constructional Steels

- Base metals, usually with chemically modified surfaces due to atmospheric elements (corrosion) and that cannot be bonded in a stress-compatible way without pretreatment.
- Mechanical surface pretreatment.

9.1.3.7 Galvanized Steels, Zinc

- Reactions with moisture, oxygen and carbon dioxide cause zinc surfaces to form consistent carbonate-alkaline and firmly adhering corrosion protection layers, which do not even flake off in the case of temperature fluctuations.
- When bonding pure zinc, its low recrystallization temperature (10–80 °C, depending on the metallurgical condition) has to be taken into account, which limits the application of heat-curing adhesives.
- In the case of zinc-plated steels, the only mechanical surface pretreatment method to be recommended is careful grinding (sponge with household cleaning powder), because of possible zinc layer damage. In the case of a damaged zinc layer, the glueline area should be protected against corrosion creep by suitable primers or by sealing of the glueline edges.

9.1.4
Adhesives for Bonded Metal Joints

Independent of the chemical structure, industrially available adhesives are characterized by the formation of strong adhesive bonds on the respectively pretreated surfaces of the materials described. This results in the criteria for adhesive selection described in Chapter 8.

The following summary should be observed in adhesive selection:

Recommended types of adhesives:

Solvent-free reactive adhesives curing at room temperature or elevated temperatures, on the basis of

- Epoxides,
- Polyurethanes,
- Methacrylates,

- Cyanoacrylates for small-area applications and limited stress,
- Anaerobic adhesives for surface sealings and thread-locking devices, furthermore,
- Hot-melt adhesives with simultaneous adherend preheating,
- Contact adhesives,
- Foamed adhesive tapes.

Adhesives not recommended:

- Solvent-based adhesives,
- Dispersion adhesives.

9.2 Plastics

9.2.1 Fundamentals

For bonding of plastics, some supplementary information on their behavior in comparison to metals is required. The essential difference is the fact that metals are generally insoluble in organic solvents. Various *plastics*, however, especially thermoplastics, are *soluble in such solvents* or at least swellable in the surface area. The result is a particular way of bonding, which is not possible with metals (Section 9.2.5).

Another difference is the fact that due to their chemical structure, clean metal surfaces enable the required adhesive forces sufficient for the bonded joint. A prerequisite for this is their good surface wetting by adhesives. This wetting behavior varies considerably according to the chemical structure of the plastics and depends heavily on the respective surface tension (Section 6.3). A typical example for poor wettability is the nonadhering inner coating of a pan with the plastic material *Teflon*. This property, in particular, led to the special application in frying and cooking, and the same property turns Teflon into the most difficult material to be bonded at all. (By the way, this plastic is not bonded into the pans in the form of film, but applied to the sand-blasted metal surface in the form of powder, which is then sintered at high temperatures). *Polyethylene* and *polypropylene* are further examples of only poorly wettable plastics.

In addition, the adherend strength is an important influence, which in many plastics accounts for only approximately 10% of the strength of metal materials. Since, due to the chemical relatedness of plastics and adhesive layers, the same or similar strength values can be assumed, butt joints are feasible, in contrast to metals (Chapter 11, 2nd rule).

9.2.2
Classification of Plastics

As described in Section 2.1.1, plastics and adhesives are very similar in their chemical structure, thus they are similarly subdivided (Figure 3.7) into the groups of

- Thermoplastics,
- Thermosetting materials,
- Elastomeres.

Table 9.1 shows the most important plastics.

Table 9.1 Most important plastics.

Thermoplastics	Abbreviation according to DIN 7728/ISO 1043	Selected trade names/ trade marks
Polyethylene	PE	Hostalen, Lupolen
Polypropylene	PP	Novolen, Hostalen-PP
Polystyrene	PS	Styron, Vestyron
Polyvinyl chloride	PVC	Vestolit, Vinnolit
Polytretrafluoroethylene	PTFE	Teflon, Hostaflon
Polymethyl methacrylate	PMMA	Plexiglas
Polycarbonate	PC	Makrolon, Merlon
Polyethylene terephthalate	PET	Vestodur, Ultradur
Polyamides	PA	Capron, Ultramid, Vestamid
Acrylonitrile-butadiene-styrene	ABS	Novodur, Terluran

Table 9.1 (continued)

Thermosetting material	Abbreviations according to DIN 7728/ISO 1043	Due to the extremely large variety of types, an indication of trade names/trademarks is not possible at this point
Phenolic resins	PF	
Urea-formaldehyde resins	UF	
Melamine resins	MF	
Unsaturated Polyester resins	UP	
Epoxy resins	EP	
Polyurethane (depending on the degree of crosslinkage, also thermoplastics and elastomers)	PUR	
Carbon-reinforced plastics	CFK	
Fiber-glass-reinforced plastics	GFK	

Elastomers/rubbers	Abbreviations according to DIN 7728/ISO 1043	Selected trade names/ trade marks
Polybutadiene	BR	Budene, Buna CB
Polychloroprene	CR	Neopren
Polyisoprene	NR	Guttapercha
Butyl rubber	IIR	Hycar-Butyl, Bayer-Butyl
Ethylene-propylene-rubber	EPM/EPDM	Buna-AP, Vistalon
Silicone	SI	Silopren
Nitrile rubber	NBR	Perbunan-N

9.2.3
Identification of Plastics

A precondition for the bonding of plastics is the knowledge of the adhesive characteristics. Trade names and product marking according to Table 9.1 help identify the adhesive; in the case of missing indications, it is difficult or almost impossible for laymen. The following two criteria may serve to distinguish between thermoplastic and thermoset material:

- Behavior at higher temperatures: This test is easily carried out by contact with a hot soldering rod. On thermoset material, the surfaces remain unchanged, on thermoplastic material a plastification, respectively, melting can be observed.
- Solubility in organic solvents: Tetrahydrofuran (THF is regarded as a solvent with very universal properties, however it is combustible and its vapors must not be inhaled (see Section 7.5.2). In addition, suitable solvents are acetone and methyl-ethyl ketone. Thermoset materials are generally insoluble, polyvinyl chloride, Plexiglas, polystyrol, rubbers and poorly crosslinked polyurethane are swellable thermoplastics.

9.2.4
Surface Pretreatment

Figure 9.1 shows a classification of plastics with regard to their bondability.

The essential criterion is therefore the respective solubility or insolubility in organic solvents. Since the surfaces of soluble plastics do not remain in their original state, a special pretreatment (except for cleaning and degreasing, if required) is not necessary. This issue is described in Section 9.2.5. Plastics insoluble in solvents are only bondable if the prerequisites *wettability* and *adhesive bond* formation are ensured. In Figure 7.2, the surface pretreatment methods available for this purpose are indicated as "physical" methods. The effect of this procedure is the chemical modification of the surface of a plastic material, especially by "integration" of oxygen molecules. This chemical modification leads to a better wetting and, at the same time, to the formation of adhesive forces. The name "physical method" refers to the fact that they utilize physical effects in the form of electrical or thermal energy. The following methods are applied:

9.2.4.1 Corona Method
Application under atmosphere at normal pressure. The Corona discharge occurs as a typically shining high-voltage discharge between two electrodes at approximately 10–20 kV and frequencies in the range of 10–30 kHz. The high energy results in the formation of oxygen atoms and ozone molecules (O_3), which have an oxidative effect on the polymer surface, thus causing an increase in surface tension and wetting behavior.

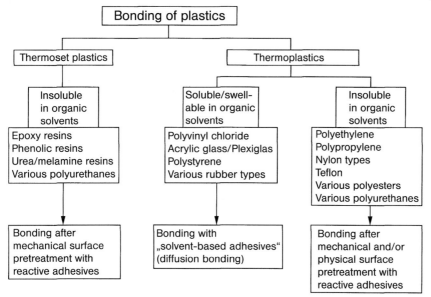

Figure 9.1 Bonding possibilities for plastics.

9.2.4.2 Low-Pressure Plasma
Operation (partly) in vacuum. Reactive gases (oxygen, hydrogen, fluorine), which are transferred into an energy-rich state (plasma) by microwave stimulation with the possibility of chemical surface modification, are fed into the plasma chamber with the adherends to be pretreated.

9.2.4.3 Atmospheric Pressure Plasma
In contrast to the Corona method, the beam generated in the plasma source shows no electrical potential. A targeted air flow conducts the focused plasma beam to the surface of the material to be treated. The treatment effect is comparable to the methods mentioned before.

9.2.4.4 Flame Treatment (Kreidl Method)
The treatment with a fuel gas-oxygen flame (propane/butane or acetylene with excess oxygen, recognizable by the blue coloration of the flame) results in a chemical and physical surface modification, also with oxidative effects. This method is particularly suitable for handycraft applications, because of its low effort and expenditures. The flame treatment time is in the range of seconds, the distance of the flame to surface should be approximately 5–10 cm. In the case of thermoplastics like polyethylene and polypropylene, care should be taken that surface melting is avoided.

9.2.4.5 Mechanical Methods

Owing to the deformability of plastics – in particular of thermoplastics – mechanical pretreatment methods are applicable only to a very limited extent. So for example, if jet pressure is too high the blasting abrasive can be "shot" into the surface. For polyethylene and polypropylene for instance, the SACO-method described in Section 7.1.2.1 has proven its worth. It develops a surface suitable for the formation of adhesive forces by means of chemically modified blasting abrasives (silication).

9.2.5
Plastics Soluble or Swellable in Organic Solvents

They include:

- Polyvinyl chloride (hard and soft),
- Acrylic glass/Plexiglas,
- ABS (acrylonitrile-butadiene-styrene-copolymers),
- Polystyrene,
- Polycarbonate and various rubber types.

A special surface treatment (except for cleaning) is not required, since the surface is partly dissolved by a solvent. As "adhesive", a suitable solvent is used in which the manufacturer has solved a certain amount of the same plastic as the plastic that is to be bonded. This bears the advantage of higher adhesive viscosity and prevents the purely low-viscosity solvent from dropping down from the adherend surfaces. The "adhesive solvent" is applied to both adherend surfaces. After a little while, the surface is swollen. Thus, the polymer molecules are more "flexible" and can combine with those of other surfaces by jamming or tangling up when fixed under pressure, that is, they "diffuse" into each other (from Latin *diffundere* = spreading, penetrate). This effect led to the name *diffusion bonding* for this kind of bonding. It is characterized by the fact that the adhesive layer has the same or a very similar composition of the plastic material to be bonded. The solvent residues remaining in the glueline evaporate afterwards through the glueline edges or through the plastic.

In Figure 9.2 the diffusion bonding process is schematically shown. The adhesive forces developing in this way are also called *autohesion*.

Solvent adhesives for polyvinyl chlorides (PVC) and products made of this material, such as pipes, molded parts, roof gutters, canvas covers, rubber dinghies, aquatic sport equipment, rain clothes and the like are of special importance. After the application of the paste-like adhesive, both adherends are immediately fixed under slight contact pressure (only applicable for plane adherends, in the case of pipe joints use rotary motions). Depending on the given temperature, setting times of 5–20 min are required. It is recommendable to observe the processing instructions of the adhesive manufacturer for these applications. On the same basis, so-called "cold-welding pastes" for PVC (floor coverings, skirting boards, etc.) are available.

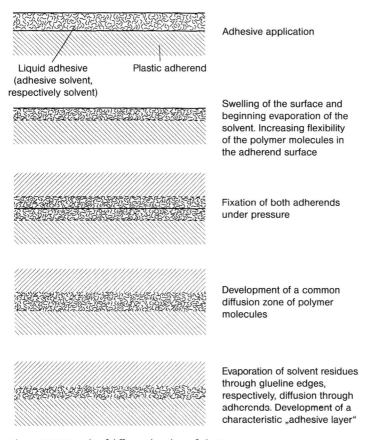

Figure 9.2 Principle of diffusion bonding of plastics.

Recommended types of adhesives:
- For the above-mentioned thermoplastics, solvent adhesives according to the diffusion bonding principle (observe indication on the packaging).
- Cold-curing reactive adhesives (hardly ever used, since the solvent adhesives described before fulfil all practical requirements).
- Cyanoacrylates.
- Polyurethanes preferably for flexible plastics, respectively, films.
- Contact adhesives.
- Radiation-curing adhesives for "crystal clear" plastics such as acrylic glass/ Plexiglas, polystyrene, polycarbonate.

Adhesives not recommended:
- Heat or hot-curing reactive adhesives, due to the limited thermal resistance of the plastic materials.
- Hot-melt adhesives, for the same reason.
- Dispersion adhesives.

9.2.6
Plastics Insoluble or not Swellable in Organic Solvents

They include:

• Polyethylene	• Teflon
• Polypropylene	• Various polyesters and polyurethanes
• Epoxy resins	• Phenol, urea, melamine resins

In such cases, the plastic surface must be made bondable by means of a surface treatment to enable bonding due to the development of adhesive forces after previous wetting. In this special case, it has become common practice to talk of *adhesive bonding* in contrast to *diffusion bonding* described in Section 9.2.5. Surface treatment may occur mechanically by brushing and sanding (apply only little pressure when sand-blasting to avoid the grains being injected into the surface). In the case of polyethylene and polypropylene, the physical methods mentioned before have to be additionally applied.

Recommended types of adhesives:

- Cold-curing epoxides.
- Polyurethanes.
- Methacrylate adhesives.
- Cyanoacrylate, in combination with a special primer also for *polyethylene* and *polypropylene*.
- Contact adhesives.

Adhesives not recommended:

- Solvent-based adhesives.
- Dispersion adhesive.

9.2.7
Plastic Foams

These materials are characterized by their porosity, although pore sizes can vary considerably. This affects both the adhesive application and the viscosity of the adhesive used. If viscosities are too low, there will be the risk of "penetration", that is, the spontaneous penetration of the liquid adhesive into the pores, rendering the thickness of the adhesive layer insufficient. In such cases, either an adhesive with higher viscosity has to be used, or adhesive application to both adherends is required twice in short time intervals.

Bonding of polystyrene foams (styrofoam) requires supplementary information. Since almost all solvent-based adhesives contain solvents capable of dissolving or swelling polystyrene, only adhesives that do not show this property, owing to their solvent composition, may be used. Therefore, special "Styrofoam" adhesives are commercially available. The dispersion adhesives described in Section 5.4 are a suitable alternative.

Recommended types of adhesives:

- Solvent-based adhesives.
- Dispersion adhesives.
- One-component polyurethane adhesives.
- Contact adhesives.
- Cyanoacrylates for small-size bonding in gel-like or paste-like composition (in the case of polystyrene, check fitness for use, because of possible foam destruction).
- Adhesive tapes.
- Hot-melt pressure-sensitive adhesives (spray application).

Adhesives not recommended:

- Epoxides, because of poorly deformable adhesive layers.
- Hot-melt adhesives.

9.2.8
Bonding of Plastics to Metals

The diffusion bonding method described in Section 9.2.5 cannot be applied to the bonding of plastics to metals, since the metal surfaces are not swellable by means of organic solvents. Bonded joints are only possible after an adequate surface treatment with the known reactive adhesives.

Attention should be paid to these material combinations, since metals and plastics expand differently under thermal stress. The ratio of the expansion coefficient is in the range of 1 : 5 (metal : plastic). In the case of small-size bonded joints, this difference is not very critical, in the case of large-size or long bonding joints, however, tensions in the adhesive layer and fractures in the bonded joint may arise at high operating temperatures (in car manufacturing, temperatures of up to 80 °C are quite usual). For this reason, reactive polyurethane hot-melt adhesives described in Section 4.2.3 are used in such cases for adhesive layers with thicknesses of several millimeters. Owing to their elastic behavior they are capable of compensating the tensions occurring in the bonded joint (see Section 10.3).

Recommended types of adhesives:

- Two-component epoxy resin adhesives for small-size bonded joints and limited thermal stress.
- Two-component polyurethane adhesives.
- Methacrylate adhesives.
- Contact adhesives.
- Foamed adhesive tapes.

Adhesives not recommended:

- Heat-curing reactive adhesives.
- Dispersion adhesives.
- Hot-melt adhesives.

For bonded joints with "crystal clear" plastics like polystyrene and acrylic glass to metals, it is also possible to apply radiation-curing adhesives (Sections 4.3.2 and 9.3.3).

9.2.9
Bonding of Plasticizer-Containing Plastics

Bonding of plastics containing *plasticizers* are of special importance. Plasticizers are added to various plastics rendering them "softer", that is, more easily deformable or flexible. An example of this is "flexible PVC" that is used for films, canvas covers and cable coverings. Plasticizers are organic compounds with relatively low molecular weight integrated in the polymers (physically), thus not linked with the macromolecules via chemical bonds. They act similar to "hinge-joints" between the molecule chains and lead to slight deformation, respectively, displacement under mechanical stress. The consequence of the fact that the plasticizer molecules are only physically "embedded", is their "migration" from the plastic into the adhesive layer (plasticizer migration) in the course of time, especially at higher temperatures. This "softening" causes the gradual failure of the adhesive layer. Thus, for the bonding of plasticizer-containing plastic materials, the application of room temperature curing reactive adhesives is preferred, since owing to the crosslinked polymer layer, they show a high resistance to penetration of plasticizers.

9.3
Glass

9.3.1
Surface Pretreatment

A characteristic property of glass is the easy binding (adsorbing) of moisture on the surface, due to their chemical structure, which may impede the development of adhesive forces. Since the usual mechanical surface pretreatment is only applicable to a limited extent, because of possible microcrack formation in the surface, and chemical pretreatment by means of etching with hydrofluoric acid, because of the high effort required, an alternative solution is surface cleaning by means of degreasing with organic solvents (ethyl alcohol, isopropyl alcohol, acetone). However, in this procedure it has to be noted that, despite degreasing, the removal of the water adsorbed on the surface is only temporarily possible. Due to the evaporation of the organic solvent used for cleaning, the glass surface cools down ("evaporation chillness") that again leads to a partly enhanced moisture adsorption in the bond area. Thus, it is recommended to heat the surface with a hair dryer after degreasing and prior to adhesive application to approximately 40–45 °C to help the adsorbed water evaporate and to enable the immediate application of the adhesive. However, the application of the adhesive to the heated surface reduces the open assembly time so that the adherends have to be fixed immediately. In addition, moderate "sanding" with a household detergent powder – and ensuing degreasing – is recommended.

9.3.2
Glass–Glass-Bonded Joints

To avoid inner tensions through thermal stress, it is recommended to use only adhesives that cure at room temperature. The adhesive selection is limited by the fact that many applications require an "invisible" glueline. In such cases, adhesives with fillers are excluded, cyanoacrylates and in particular radiation-curing products (Section 9.3.3) are the suitable choice. If the visual appearance of the bonded joint is not important, two-component reactive adhesives based on expoxides, polyurethanes, methacrylates, contact adhesives and, if required, adhesive tapes are recommended.

Adhesives not recommended:

- Hot-melt adhesives, because of the high temperature of the melt and very low thermal conductivity of glass (risk of fracture).
- Solvent-based adhesives.
- Dispersion adhesives.

In many cases, glass and even porcelain bondings are carried out as repairs in households. In such cases, it has to be pointed out that, compared to the two-component reactive adhesives, in particular epoxides, cyanoacrylates show only a limited bond strength in the face of high temperature and humidity stress in connection with rinsing agents in the cleaning equipment.

9.3.3
Bonded Glass Joints with Radiation-Curing Adhesives

When radiation-curing adhesives (Section 4.3.2) are applied, care has to be taken that – regarding the radiation spectrum – the adhesive used and the UV radiation source have to be adapted to each other. The photoinitiator contained in the UV adhesive requires specific wavelengths for the adhesive to be able to cure and develop a constantly strong and stress-related adhesive layer. It is recommendable to buy adhesive and radiation source in the form of a "system" from one supplier. Moreover, it has to be mentioned that "transparency of glass" in the visible region does not necessarily result in sufficient UV transmittance. The radiation energy reaching the adhesive layer can vary considerably depending on the glass composition and thickness of the material. Therefore, at any rate, the UV transmittance

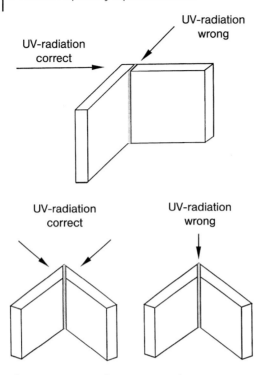

Figure 9.3 Correct and incorrect UV-radiation curing.

of the glass has to be verified by a UV sensor. Since the adhesive layer applied absorbs the UV-radiation, "thicker" adhesive layers are provided with increasingly lower radiation energy, with the result of uneven curing of the adhesive layer. For this reason, the arrangement of emitter and adhesive layer, depicted in Figure 9.3, has to be observed.

9.3.4
Glass–Metal Joints

The major principles to be observed with regard to such material combinations have already been discussed in Section 9.2.8 on plastic–metal joints. Even in case of the adherend combination "metal–glass", the different expansion coefficients of both materials have to be considered. Metals like aluminum, steel, copper and brass expand twice as much as normal glasses. The thermal expansion of adhesive layers of epoxy resins, phenol resins or acrylates is approximately ten times higher than that of glasses. In order to avoid tensions in the bonded joint, the binding adhesive layers should show sufficient deformability (polyurethanes, special flexibly adjusted epoxy resin adhesives) and must by no means be stiff and brittle. To avoid unnecessary thermal stress already during the production of the bonded joint, it is advisable to use only cold curing adhesive systems. For glass bonds exposed to high stress due to temperature fluctuations and UV radiation (e.g., windows glued into the car body) manufacturers offer special primers to be applied to the glass surface prior to adhesive application.

9.4
Rubber Products

The term *rubber* is the common name used to denote natural and synthetic rubber in its vulcanized form. Depending on the degree of crosslinkage, a distinction is made between soft and hard rubber. The original name of crude, unvulcanized rubber is *caoutchouc* and derives from the Indian words *caa* = tears and *ochu* = tree or even *cahuchu* = crying tree, referring to the dripping out of latex after cutting the bark of rubber trees (hevea brasiliensis).

The different rubber types and products show great elasticity. Therefore, the joining adhesive layers must be able to deform in the same way when exposed to stress. Particularly suitable adhesives are solvent-based adhesives with rubber portions ("rubber solvents") and contact adhesives. In the first case, the adhesive process is based on the diffusion bonding principle described in Section 9.2.5. Polyurethane and cyanoacrylates are particularly suitable as reactive adhesives. The latter are used for butt joints in the manufacturing of rubber rings for sealing purposes. For such applications, equipment with rubber profiles of different diameters, cutting devices, adhesive and dimensions are commercially available. The EPDM "rubber" widely used in car manufacturing, because of its excellent ageing resistance (belongs to the group of thermoplastic elastomers) can be bonded only to a limited extent due to its chemical composition and the various formulations. Cyanoacrylates are recommended as adhesive; at any rate, respective tests are required since not all EPDM types show the same bondability.

Because of the metals' insolubility in solvents, *rubber–metal joints* can only be produced with solvent-free reactive adhesives or even contact adhesives after suitable pretreatment. In the case of rubber–metal joints used as shock-absorber, respectively, oscillating elements in car manufacturing, the development of the bonded joint occurs during vulcanization. The term *vulcanization* dates back to the method developed by Goodyear around 1840 for the crosslinking of natural rubber with the simultaneous impact of sulfur and heat, which were known as byproducts of the "volcanism".

Since different rubber qualities often show components able to diffuse into the surface, or the surfaces of which are coated with processing material (e.g., talcum), mechanical surface roughening is recommended or – if possible – the straight cutting off of the external layer (approximately 0.1–0.2 mm) of, for example, rubber profiles (with razor blades or sharp knives, cuts will be straighter than with scissors).

9.5
Wood and Wood Products

As a natural product, wood is characterized by a more or less distinctive porous structure. Furthermore, the possible moisture content of wood with its effects on the applied adhesive and its curing behavior have to be noted:

- In the case of dispersion adhesives, excessive wood moisture may delay the penetration of water from the dispersion into the material, and thus extend the setting time.
- In the case of condensation adhesives, which cure under separation of water, this may be entrapped in the adhesive layer with ensuing shrinkage symptoms.
- In the case of hot-melt adhesives, high moisture content resulting from the high temperature of the melt can lead to the formation of water vapor during the wetting process, thus reducing the adhesive forces.

Practical experience shows that wood moisture must not exceed a value of 8–10%. Adhesives are selected according to the type of joint to be bonded and the kind of stress:

- The bonded joints are defined as area bondings, joint bondings and assembly bondings. Characteristic features of the adhesive to be used are its open assembly time, which has to be considerably longer for area bondings (veneers, high-pressure laminate) than, for example, for the fixing of dowels or the assembly of corner joints. While in the first case, the application of heat-curing adhesives occurs by heated presses, the production of assembly joints is possible with hot-melt adhesives or quickly setting dispersions.
- The *kinds of stress* are described by the conditions of the climatic effect to be expected, in particular of moisture.

Recommended types of adhesives:

- Dispersion adhesives for small-size and area bonded joints.
- Hot-melt adhesives for small-size bonded joints (owing to short open assembly time).
- Contact adhesives.
- Solvent-based adhesives.
- Polycondensation adhesives (usually for industrial application) on the basis of phenol, resorcinol, urea and melamine formaldehyde resins. For the processing of such adhesives used, in particular, for wood constructions in moisture-loaded buildings, such as swimming pools, heatable presses are required.
- Polyurethane adhesives.

So-called "white glues" are particularly important. The chemical basis is polyvinyl acetate (PVA), the bonded joints – or traditionally *adhesive gluings* – are characterized by very high adhesive strengths, which usually result in adherend fractions after destructive testing. In order to avoid possible tensions, it is recommended to let the bonded joint set under even pressure (clamp) for more than the indicated time interval. Contact adhesives are particularly suitable for the gluing of plastic coatings (urea-, melamine resin) and also for the combinations wood-leather, textiles, veneers.

9.6
Porous Materials

This area comprises, in particular, the material groups of

a) Paper, paperboard, cardboard, photos.
b) Wood, plywood, balsa wood, chip board, veneers, cork.
c) Textiles, felt, leather (as typical representatives of flexible materials).
d) Porcelain, ceramics, baked clay, concrete, marble, artificial stones.
e) Plastic foams (see Section 9.2.7).

The special adhesive behavior of these materials is to be seen in the respective surface structure, which renders the positive adhesion described in Section 6.1 particularly important.

When choosing the adhesive, the following material properties have to be considered:

- Surface structure, that is, pore size as well as surface roughness.
- Deformation behavior, flexibility.

Therefore, the adhesives are to be chosen according to the criteria:

- Viscosity.
- Deformability of the adhesive layer.

Recommended types of adhesives:

Group a): Solvent-based adhesives, hot-melt adhesives (paperboards, cardboards), dispersion adhesives (for further information refer to Section 5.2).
Tip: For bonding of papers, paperboards, photos and the like, glue-sticks – available for detatchable and nondetachable bonded joints – are recommended because of their convenient application (Section 5.8).

Group b): Solvent-based adhesives, dispersions, hot-melt adhesives, contact adhesives (for further information refer to Section 9.5).

Group c): Hot-melt adhesive films (heat sealing), contact adhesives, solvent-based adhesives.

Group d): Solvent-based adhesives, two-component epoxy resin adhesives, two-component methacrylates for assembly purposes (priming, filling, smoothing, sealing, etc.) For these applications, a multitude of so-called *building/construction adhesives* is commercially available.

Group e): See Section 9.2.7.

Complementary literature to Chapter 9:
[B5, P2, T1].

10
Strength, Calculation and Testing of Bonded Joints

10.1
The Term "Strength"

What do we understand by *strength* of a material or a bonded joint? Knowingly or unknowingly, we distinguish materials according to their strength behavior. Steel, for example is referred to as "solid", because it can be loaded by forces, for example, a steel cable by tensile load. Plastics like polyethylene or Plexiglas, on the other hand, are regarded as less solid; rubber for instance is described as "soft". A material is said to be solid when it is hardly or not at all deformable under the influence of external forces. Less solid materials show a visible deformation (e.g., a plastic cable), and "soft" materials like rubber can be deformed (expanded) to a large extent with little effort.

The essential reason for these different appearances is the "internal strength" described in Section 6.4, the "cohesion" of the materials, due to their differing atomic or molecule structure.

The following example serves to explain "strength": A wire with a cross-section of 1 square millimeter (mm^2) is clamped into a tension testing machine and torn apart (Figure 10.1).

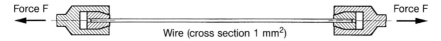

Figure 10.1 Tensile test.

Here, a certain force *F* is required. In the second test, a wire with a cross-section of 2 mm^2 is torn apart, which requires twice the force, thus 2 *F*. But is this wire therefore twice as strong as the wire with a cross-section of 1 mm^2? This is certainly not the case, since twice the force is only required, because the cross-section is twice as large. It is clear to see that the force required and the strength are not the same. The force necessary for tearing the sample apart depends on the cross-section of the latter.

Applied Adhesive Bonding: A Practical Guide for Flawless Results. Gerd Habenicht
Copyright © 2009 WILEY-VCH Verlag GmbH & Co. KGaA, Weinheim
ISBN: 978-3-527-32014-1

According to an international agreement, force is measured in a determined unit, called the "Newton", in remembrance of the English physicist Isaac Newton (1643–1727). The abbreviation is "N".

When tearing apart the 1 mm² wire, a force of

$$240 \text{ N}$$

is reported to have been measured at the tensile testing machine, in the case of the 2 mm² wire, a force of

$$480 \text{ N}$$

is said to have been measured. To allow a comparison of these two values, they must refer to the same cross-section; according to the agreement, this is the value of 1 mm². Consequently, the force 480 N, measured for the wire with 2 mm², must be divided by the number 2:

$$\frac{480 \text{ N}}{2} = 240 \text{ N}$$

Referring to the same cross-section, both wires require the same force to be torn apart. From this, the term "strength" can be deduced. It represents the tensile strength at break required for the cross-section of a material of 1 mm²:

$$\text{Strength of a material} = \frac{\text{tensile strength at break required}}{\text{area in mm}^2} \left[\frac{\text{N}}{\text{mm}^2}\right]$$

Thus, both wires tested have the same strength of

$$\frac{240 \text{ N}}{1 \text{ mm}^2} = 240 \frac{\text{N}}{\text{mm}^2}$$

respectively,

$$\frac{480 \text{ N}}{2 \text{ mm}^2} = 240 \frac{\text{N}}{\text{mm}^2}$$

Thus, independent of the dimensions of the test pieces, the strength values obtained are always comparable. These relationships refer also to bonded joints for which the term "adhesive strength" has become established for destructive testing of overlapped test pieces (Figure 10.2). It is abbreviated by the Greek letter τ (pronounced *tau*) referenced with the index B:

$$\text{Adhesive strength } \tau_B = \frac{\text{max. force } F_{max} \text{ at break}}{\text{adherend surface } A} = \frac{F_{max}}{A} = \frac{\text{N}}{\text{mm}^2}$$

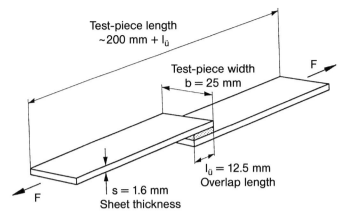

Figure 10.2 Single-lap test piece according to DIN EN 1465.

Comment: The dimension for the strength, respectively, adhesive strength given in N/mm², that is, the force required to load a material or a bonded joint with a specific area until it breaks, is easy to understand from a didactical point of view. In the course of the rearrangement, respectively, amendment of standards, the dimension

MPa (Megapascal)

has also been introduced in bonding technique. It is originally deduced from the unit used for pressure acting on a surface

1 Pascal (1 Pa) = 1 Newton per squaremeter = 1 N/m²

It is named after the French physicist Blaise Pascal (1623–1662). So, *inter alia*, air pressure is indicated in hectopascal (hPa = 10^2 Pa, former 1 atmosphere atm = 1013,25 hPa). Basis is the Law on Units in Metrology of the year 1970. This unit is applicable to each area-related or cross-section-related force. From the relationship

1 Pa = 1 N/m² = 1 N/1000 × 1000 mm²
1 000 000 Pa = 1 N/mm²

which results in

1 MPa = 1 N/mm²
(M = Mega for a millionfold = 10^6)

thus, strength values in N/mm² often cited in the literature correspond to the values given in MPa.

10.2
Test Methods

The most important tests on bonded joints are targeted at the determination of the strength under precisely defined conditions. In order to obtain comparable results from such tests on different test stations, for example, at the adhesive manufacturer and the adhesive user, the test conditions have to be stipulated in detail and must be binding. For this purpose, test standards have been issued by the German Institute for Standardization (DIN) and the European Standards (EN) in cooperation with interested technical groups. The standards for tests in the field of adhesive technology, for example, contain indications regarding material and dimensions of test pieces, the test method to be applied (test equipment, test speed), if required even surface pretreatment of test pieces and other test criteria to be taken into account.

The continuously advancing technological developments inevitably lead to changes in these rules and regulations. The latest issues of the standards can be inquired at

Beuth-Verlag GmbH
Burggrafenstrasse 6, 10772 Berlin
www.beuth.de or www.cenorm.be

The test methods are subdivided into two categories, *destructive* and *nondestructive methods*. One of the latter, the "ultrasonic method", is applied in special fields of adhesive technology, for example, in aircraft construction. It is based on the fact that sound waves in test pieces propagate differently, in dependence on the evenness of their structure, which can be disturbed by the presence of pores or imperfections, thus enabling the recording of failures.

The most important methods, however, are the destructive test methods by means of special test pieces which allow for the determination of the strength of bonded joints.

10.2.1
Adhesive Strength Testing

This test is carried out according to the standard DIN EN 1465 *Adhesives – Determination of the tensile shear strength of high strength lap joints*. The test piece has the dimensions according to Figure 10.2.

Tensile shear strength (adhesive strength), in the sense of this standard, is defined as the maximum force F_{max} at the break of the bonded joint in relation to adherend surface A. The adherend surface A results from the test piece width b (25 mm) and the overlap length $l_{ü}$ (12.5 mm):

$$A = b \times l_{ü} = 25 \times 12.5 = 312.5 \text{ mm}^2$$

Calculation example:

What is the adhesive strength of a single-lap bonded joint with the above dimensions, if the testing machine measures a maximum force of 7500 N at the break of the bonded joint?

Solution:

$$\tau_B = \frac{F_{max}}{A} = \frac{F_{max}}{b \times l_{ü}} = \frac{7500}{25 \times 12} = 24 \; \frac{N}{mm^2}$$

This example is an extremely simplified depiction of the actual situation, nevertheless the following reference values of bond strengths determined according to this method can be assigned to the adhesives given below:

	Bond Strength τ_B MPa (N/mm^2)
• Heat/hot cured, strongly crosslinked polyaddition adhesives (epoxy/phenol resins)	25–35
• Epoxy resin adhesives cured at room temperature	20–30
• Polymerization adhesive cured at room temperature (methacrylate, cyanoacrylate)	10–20
• Reactive polyurethane hot-melt adhesives	5–10
• Hot-melt adhesives (thermoplastics)	10–15

However, for reasons described in the following section, these data are suitable for constructive calculations only to a limited extent.

10.2.2
Tensions in Single-Lap Bonded Joints

Due to the following reasons, the determination of "pure" strength values, characteristic of the adhesive layer only, is not possible with the test piece dimensions according to DIN EN 1465 shown in Figure 10.2, because (see Figure 10.3):

- Force F is eccentrically applied, which results in a deformation (bending) of the adherends in the area of the two overlapping ends.
- This results in increased tensile stress σ_z (normal stresses, peel stresses) for the adhesives layer in these areas caused by the bending moment M_b.
- The adherend extension occuring in the area of the overlap ends due to the application of loads results in tensile and shear stresses parallel to the adhesive layer (τ_ε).
- The adherend displacements parallel to the adhesive layer create shear stresses (τ_v).

The interfering stresses described before are the reason why this glueline geometry does not enable the determination of mere shear stresses as a basis for constructive calculations. The measured "strength value" is a complex value, which includes

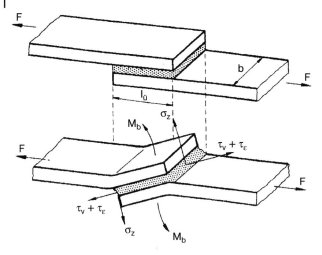

Figure 10.3 Tensile shear stress of a single-lap bonded joint.

both adherends and adhesive layer properties. The term "tensile shear strength" determined as standard for "adhesive strength" is based on the tensile and shear stresses simultaneously occuring in this test method.

Depending on the strength of the adherend materials, their thickness and their overlap length, the reacting force leads to high stress peaks at the overlap ends, thus adhesive layer area is exposed to correspondingly high stresses. This may result in crack formation that will then – in the stress direction – continue from both sides of the bonded joint to the center of the adhesive layer until it breaks. Thus the "load behavior" of the bond is, in consideration of the overlap length, "inhomogeneous". This fact leads to the statement that an increase in the overlap length at constant adherend width does not enable the transfer of proportionally higher forces. Knowing the adherend thickness, it is possible to dimension roughly the required overlap length $l_ü$ of metal or even high-strength nonmetal materials according to the relationship

$$l_ü \approx 10 \times s$$

Based on the given relationship, a steel sheet with a thickness of 0.8 mm would have an overlap length value of $l_o \approx 8$ mm. With an adherend width of 25 mm and a bond strength of the adhesive of $\tau_B = 20$ N/mm² a force of

$$F = \tau_B \times b \times l_ü = 20 \times 25 \times 8 = 4000 \text{ N}$$

can be transferred.

However, it has to be pointed out again that, in this way, only approximate results regarding adherend geometries and forces to be transferred can be calculated. A very precise calculation method considering the complex parameters "adhesive", "adherend material", "glueline geometry" as well as other parameters is the finite-element method (FEM), which requires high expenditures regarding computer-aided calculation approaches. Further determining factors to be taken into account are *reduction factors* for dynamic and environmental stresses as well as for long-term resistances.

10.2.3
Shear Strength Testing

To enable the determination of almost "pure" strength values for the adhesive layer, the parameters "eccentric application of load" and "adherend extension/deformation" must be eliminated. This is the case in the test piece geometry depicted in Figure 10.4 according to the standard ISO 11003-2 "Shear testing method for thick adherends".

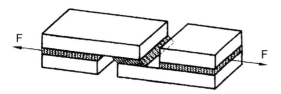

Figure 10.4 Shear stress of a single-lap bonded joint.

The centrical force transmission is obtained by gluing sections of the same adherend thickness onto the adherends in the force transmission area. The adherend extension will be eliminated by a larger adherend thickness (5 mm instead of 1.6 mm) and a reduced overlap length. Due to the homogeneous stress distribution, the respective test results are thus based on defined shear stress conditions. Adhesive layer values, to a large extent independent of the adherend, such as shear modulus, shear strength and deformation behavior, are available as a basis for precise calculations.

10.2.4
Peel Resistance Testing

The testing of the peel resistance occurs according to the standard DIN EN 1464 *Adhesives – determination of peel resistance of high-strength bonded joints – climbing drum peel test* (respectively, T-peel test) and serves the determination of the resistance of bonded metal joints to peeling forces (Figure 10.5).

When loading this test piece with force *F* it is noticed that, unlike in the tensile shear test, the force is not applied to an area $A (= b \times l_{ü})$, but to a line X…X. The other area of the adhesive layer remains unstressed. Thus, in this case "strength" cannot be defined as "force per area", but the force referring to a line is called *peel resistance*. If the test piece shown in Figure 10.5 is torn apart by means of force *F*, and the force over the peeled distance is recorded, the following peel diagram results (Figure 10.6).

10 Strength, Calculation and Testing of Bonded Joints

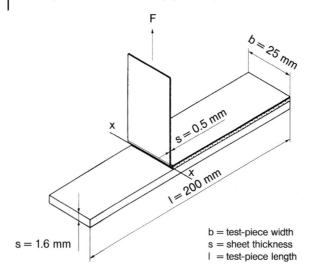

Figure 10.5 Test piece for peel test according to DIN EN 1464.

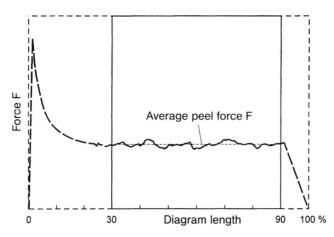

Figure 10.6 Peel diagram.

Due to the occurring fluctuations, only the range of 30–90% of the peeled test piece length will be utilized in the analysis for the determination of the average peel force. Then, with reference to the test piece width b, the peel strength p_s results

$$p_s = \frac{\overline{F}}{b}$$

Calculation example:

What is the peel strength of a bonded test piece with the test piece width $b = 30$ mm at a measured average peel force $F = 22.5$ N?

Solution:

$$\text{Peel strength } p_s = \frac{\text{average peel force}}{\text{test piece width}} = \frac{\overline{F}}{b} = \frac{22.5}{30} = 0.75 \frac{N}{mm} = 7.5 \frac{N}{cm}$$

This test method is preferably used for the comparative evaluation of adhesives and surface pretreatment methods, since it enables the indication of the differences in the adhesive and cohesive behavior of the adhesive layers with high sensitivity. Pressure-sensitive adhesives (adhesive tapes, adhesive labels) are also tested according to this principle.

After the breakage of the test piece, it is not only interesting to know the average peel force for the calculation of the peel ressitance; it is just as important to determine the causes of the break in the bonded joint. If possible, a fracture analysis is carried out for this purpose by means of a microscope or magnifying glass. There are three different possibilities for fractures: adhesion, cohesion or mixed fracture (Figure 10.7).

In most cases, adhesion fractures are a sign of insufficient surface pretreatment. In the `case of mixed fractures, the reason is to be seen, with high probability, in an incomplete or uneven degreasing. The neglected degreasing after blasting can also be seen as a cause, if the compressed air has not been absolutely fat-free. A cohesion fracture may be the consequence of an adhesive layer not completely cured or of the application of an inadequate adhesive mix.

Section 7.4 gives details on the most important reasons for failures and the respective remedying measures.

10.2.5
Test Methods for Short-Term and Long-Term Stresses

Only very rarely are bonded joints stressed exclusively under normal conditions. For this reasons, testing under the environmental effects of temperature, natural and artificial climates are required. Ageing tests are usually carried out on standardized test pieces, mainly on single-lap bonded joints exposed to respective environmental conditions and then tested according to the provisions of the respective standards.

The short-time tests usually applied may stress the bonded joints as follows:

- Warm, humid conditions, with defined moisture concentration (accumulation of water, respectively, relative humidity) and defined temperature. The results are swelling processes in the adhesive layer.
- Corrosive media that cause a particularly strong attack and thus damage in the bonded joint within relatively short time. The best-known test method is the salt spray test.
- Temperature changes are carried out with a simultaneous moisture influence in the range of –40 to 80 °C. They stress the bonded joint by expansion and contraction of the water diffused into the adhesive, respectively, boundary layer.

The most important standards are:

DIN EN ISO 9142	Adhesives – Guide to the selection of standard laboratory ageing conditions for testing bonded joints
DIN EN ISO 10365	Adhesives – Designation of main failure patterns
ISO 14615	Adhesives – Durability of structural adhesive joints – Exposure to humidity and temperature under stress
DIN 50021	Spray tests with different sodium chloride solutions
DIN 5328	Testing of metal adhesives and metal joints – Conditions for testing at different temperatures
DIN 54456	Testing of structural adhesives – Test of resistance to climatic conditions.

Apart from the tests defined in the standards, a number of in-house test methods exist.

Independent of the test methods applied and the respective results, a precise analysis of the failure source is required. Here, "fracture-type time-diagrams" are used in which an allocation of failure sources is made over the test period (adhesion fracture, cohesion fracture, adherend fracture, corrosion, q.v. Figure 7.8 and Figure 10.7).

A measure of the damages affecting the bonded joints is the loss of strength of the aged test pieces in contrast to the nonaged test pieces, which may be indicated in the form of reduction factors (Section 10.2.2). Only property testing of bonded joints under these complex stresses composed of mechanical and environmental influences enables an extensive statement on the behavior in practical application.

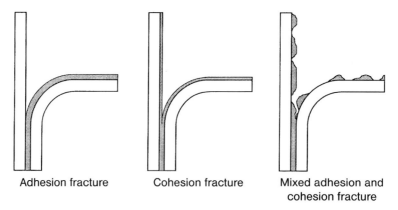

Figure 10.7 Fracture types of bonded joints.

Since results of long-term tests under original conditions cannot be relied on due to frequently short development and production cycles, tests in temporally nested form are required to enable respective information on long-term behavior. This forces a tightening of the test conditions, often with the consequence that based on the test results cause and effect cannot be exactly defined. Thus short-term tests, the basis for the assessment of long-term behavior, are always a compromise between a minimum time effort and a test result coming as close as possible to the behavior in practice. Therefore, conclusions drawn from tightened short-term tests to long-term behavior of bonded joints require very critical consideration. The following comparison might underline this problem metaphorically: "If an egg is exposed to a temperature of 100 °C for a short time (5 min), it will turn into a breakfast egg. If an egg is exposed to a temperature of only 37 °C for a long time (28 days) a fledgling will develop" (Univ.-Prof. Dr. Ing. Klaus Dilger, Institute for Bonding and Welding Technology, Technical University of Brunswick).

10.3
Elastic Bonding

The reactive adhesives described in Chapter 4, based on epoxides, phenol resins and acrylates, are usually applied in adhesive layer thicknesses of 0.05–0.2 mm (Section 7.2.3.3). Owing to their crosslink density, high bond strength values of up to 35 MPa are achieved, the respective adhesive layers, however, show only poor deformation properties. High stresses by shear and peel forces may lead to crack formation at overlap ends due to tension peaks and consequently to a failure of the bonded joint (Section 10.2.2). Experience shows that in the case of adhesive constructions with similar materials and respective glueline dimensions, the safety of the construction is nevertheless guaranteed. Special conditions, however, are given in the case of adherend combinations with materials of different thermal expansion coefficients in changing temperature ranges.

In these cases, the adhesive layers must show sufficient deformability that is guaranteed, in particular, by polyurethane adhesives. They provide the prerequisite for "elastic bonding" in car manufacturing with various material combinations of steel, aluminum, plastics and glass. Furthermore, the adhesive layer thicknesses are in the range of millimeters, thus occurring stresses can be relieved.

An example of the application of elastic bonding shall be the following simplified calculation (without consideration of the temperature dependence of the thermal expansion coefficient) regarding the stress of a GRP roof acting on the steel structure of a bus (from the book "Elastic Bonding"; see Reference B6).

Length of the glueline L_o	=	8000 mm
Thermal expansion coefficient steel α_{st}	=	$12 \cdot 10^{-6}$ K^{-1}
Thermal expansion coefficient GFK α_{GFK}	=	$20 \cdot 10^{-6}$ K^{-1}
Temperature difference (summer setting 90 °C – 20 °C) ΔT	=	70 K

For thermal expansion it generally applies:

$$\Delta L = L_0 \, \alpha \, \Delta T$$

GRP-roof	$\Delta L = 8000 \times 20 \times 10^{-6} \times 70 \;=$	11.2 mm
Steel structure	$\Delta L = 8000 \times 12 \times 10^{-6} \times 70 \;=$	6.7 mm
Difference of length alternations		4.5 mm

Thus, in the case of the bonded roof able to be displaced at both ends, an alternation of length of 2.25 mm occurs at each end. The adhesive layer thickness is usually dimensioned in the same size as the total alternation of length, in this case at least 4.5 mm. Consequently, at the overlap ends the adhesive layer will be stressed to a maximum shear of 50%.

In contrast to the relationships described in Section 10.2, the deformability of the adhesive layer and the related homogeneous stress distribution enable a simplified calculation of bonded constructions. Due the stress peaks at the overlap ends being omitted to a large extent, there is an approximate proportionality between overlap length and acting force.

A prominent example for the possibilities of elastic bonding is provided by the bonded joint shown in Figure 10.8. Modern manufacturing methods increasingly comprise modular structures of system components supplied almost ready-to-operate. In rail vehicle manufacturing, for example, complete driver's cabs including operating panel, side and front windows are connected to the frame structure by bonding. The glueline (black line), which connects the driver's cab of the suburban train with the carriage bodywork, is clearly visible.

Figure 10.8 Bonded driver's cab – carriage bodywork of a suburban train.

10.4
Shaft-to-Hub Joints

The following illustration (Figure 10.9) shall explain the basic coherences for the dimensioning of a bonded shaft-to-hub joint. Here, a limitation to the essential geometrical and mechanical parameters is required, since the reduction factors, viscoelastic adhesive layer behavior, stress development, surface geometry of the adherends and the like, do not allow for a detailed consideration at this point (Figure 10.9).

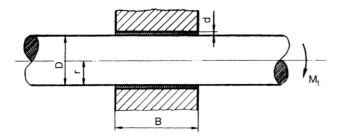

Figure 10.9 Bonded shaft-to-hub joint.

The calculation of shaft-to-hub bonded joints is based on the following parameters:

For adhesive strength, it generally applies

$$\tau_B = \frac{F}{A}$$

(F = acting force, A = area)

$$A = 2\pi \frac{D}{2} \times B$$

(D = hub diameter, B = shaft width)

$$M_t = F r = F \frac{D}{2}$$

(M_t = torsional moment)

With the adhesive strength τ_B being replaced by the torsional shear strength τ_T of the adhesive, the torsional moment to be transmitted results

$$M_t = \tau_T \times \frac{\pi D^2 B}{2}$$

Calculation example:

- Torsional moment to be transferred M_t = 600 Nm
- Hub diameter D = 30 mm
- Torsional shear strength adhesive τ_T = 20 MPa

Thus, the shaft width B to be chosen is calculated as

$$B = \frac{2 \times 600 \times 1000}{20 \times 30^2 \times \pi} = 21.2 \text{ mm}$$

Complementary literature to Chapter 10:
[B5, B6, G1, K2, K3, M3].

11
Constructive Design of Bonded Joints

If bonded joints have to "last" when stressed by forces, the choice of the adequate adhesive is just as important as the "bonding-compatible" arrangement of the adherends in the glueline. Here, some basic rules have to be observed:

Rule 1

Bonded joints have to be designed in a way that forces applied cannot result in peeling or cleavage in the adhesive layer (Figure 11.1).

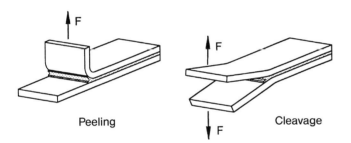

Figure 11.1 Peel and cleavage stress in a bonded joint.

As already described in Section 10.2.4 on the determination of peel resistance, the reason is that in this case, only line stress instead of surface stress acts on the adhesive layer, to which it is very sensitive. Then forces cannot be transferred. A well-known application of this effect is the tearing-off of an adhesive tape from skin. Due to the tape's slow "peeling-off" in a very small angle, only a minimum force transfer to the skin occurs and pain perception is low. The same applies to the tearing-off of an adhesive label from a substrate while, depending on the kind of pressure-sensitive adhesive, the nondestructive peeling-off of the paper label is possible although its particular strength is very low. A simple test allows for the depiction of a force transfer in the case of peel stress, which is much lower compared to that of shear stress (Figure 11.2).

Applied Adhesive Bonding: A Practical Guide for Flawless Results. Gerd Habenicht
Copyright © 2009 WILEY-VCH Verlag GmbH & Co. KGaA, Weinheim
ISBN: 978-3-527-32014-1

11 Constructive Design of Bonded Joints

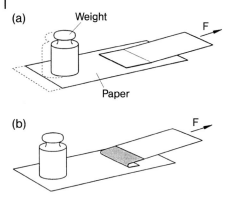

Figure 11.2 Force transfer in the case of shear and peel stress.

A weight G (approx. 250 g) acts on a sheet of paper (on a smooth board). A paper coated with pressure-sensitive adhesive (Multi-adhesive memo sheets available at the stationery shop are best suited) is loaded by force F, according to Figure 11.2, drawing (a) (shear stress). It is possible to draw the paper sheet over the board with the weight on it.

In the case of drawing (b), that is, peel stress (the memo sheet is turned and bonded in an angle of 180°) the paper with the weight on it remains in its position, force transfer is not possible.

From a design-engineering point of view, peel stress can be avoided by the possibilities shown in Figure 11.3.

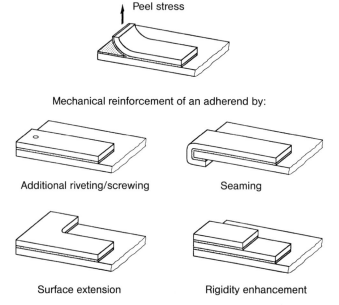

Figure 11.3 Design possibilities to avoid peel stress.

Rule 2

In order to enable the transfer of forces by means of a bonded joint, a sufficient adhesive surface must be provided between the adherends. This requirement is emphasized by Figure 11.4:

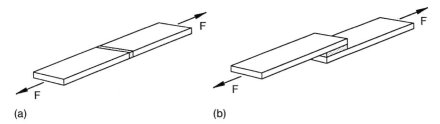

Figure 11.4 (a) Tensile stress and (b) tensile shear stress.

In the case of metal adherends with their high strength, the adhesive layer is the "weakest link" in the "strength chain" under tensile stress in case (a). Stressed by force F, the adhesive layer in such a bonded joint will break. In case (b), under tensile shear stress, the adhesive layer can be expanded to a certain extent (see Section 10.2.2) by enlarging the overlap length $l_{ü}$, and a bigger force F can be transmitted. In general, bonded joints require sufficient adherend surfaces.

Rule 3

Rule 2 is not always applicable to bonding of plastics. As mentioned in Section 2.1.1, cured adhesive layers can be compared to plastics as far as their strength is concerned. Then, under tensile stress, the adhesive layer would not be the weakest link in case (a) of Figure 11.4, since adherends and adhesive layer have comparable strengths. Thus, such butt joints are possible and also common for bonded plastic joints.

Rule 4

This rule refers to circular bonded joints, for example, tubular bonded joints or shaft–hub joints, with heat or hot-curing reactive adhesives, if different metal materials are to be bonded, Figure 11.5.

As is generally known, materials expand more or less with increasing temperature. If, for example, the (inner) tube 1 of the tubular bonded joint shown in Figure 11.5 expands more compared to the (outer) tube 2 during heat or hot curing of the adhesive, this will result in a reduction of the glueline gap and the adhesive, initially liquid, will be squeezed out of the glueline. After cooling down, defects in the adherend surface occur. In shaft–hub joints, this occurs in a similar way.

11 Constructive Design of Bonded Joints

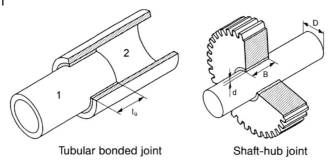

Tubular bonded joint   Shaft-hub joint

Figure 11.5 Geometry of tubular bonded joints.

Consequently, the material with the greater expansion should always be the *outer* adherend (tube 2, respectively, the hub). Cold-curing adhesives are not affected by this problem, and this is the essential reason for using such anaerobic adhesives described in Section 4.3.4 for such applications. In order to avoid the adhesive's being "displaced" during the joining of the adherends, at least one of them should be chamfered in an angle of 15–30° and the adherends should be joined under slow circular motions, Figure 11.6.

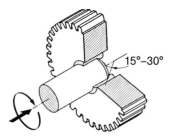

Figure 11.6 Shaft–hub bonded joint with chamfered shaft.

In conclusion, the following principles apply to the constructive design of bonded joints:

- Butt joints are unfavorable for force transmission in the case of metal materials; adhesive designs with shear stress acting on the adherend surface (overlap joints) are to be preferred.
- Because of their linear impact, peel or crack stresses of bonded joints have to be avoided at any rate.
- In general, bonded joints shall provide sufficient adherend surfaces.

Figures 11.7 (a–d) show some additional examples of favorable and unfavorable arrangements.

Complementary literature to Chapter 11:
[A1, F2, L3].

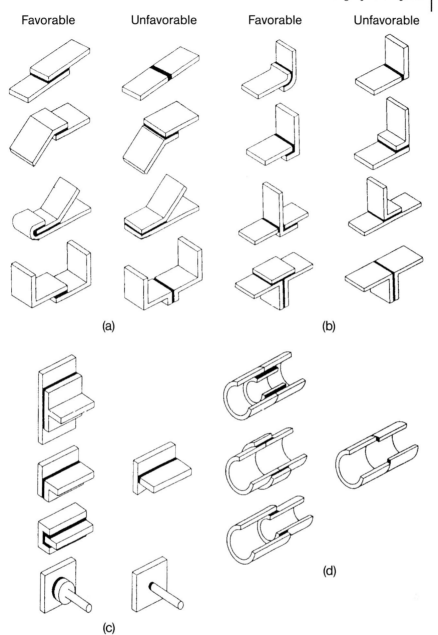

Figure 11.7 (a) Plane bonded joints; (b) corner joints; (c) attached joints; (d) tubular bonded joints.

12
References

The following references are deliberately limited to specialist books, since – in contrast to publications of technical and scientific magazines – they are easily available to the interested reader. The mentioned specialist books provide a comprehensive possibility to get informed about individual topics.

A1 Adams, R. D.; Comyn, J.; Wake, W. C.: Structural Adhesive Joints in Engineering. Chapmann & Hall, London, 1997 (ISBN 0-412-70920-1).

A2 Adams, R. D. (Ed.): Adhesive Bounding – Science, Technology, Applications. Woodhead Publishing Ltd., Cambridge, UK, 2005 (ISBN 1-85573-741-8).

B1 Benedek, J.: Development and Manufacture of Pressure – sensitive Products. Marcel Dekker Inc., New York, 1998.

B2 Benedek, J.; Heymans, L. J.: Pressure – Sensitive Adhesives Technology. Marcel Dekker Inc., New York, 1997 (ISBN 0-8247-9765-5).

B3 Benedek, J.: Pressure-Sensitive Design. Theoretical Aspects, Formulation, Application. Vol. 1, Brill Academic Publishers, Leiden, The Netherlands, 2006 (ISBN 9067-6443-90).

B4 Blackley, D. C.: Polymer Latices Science and Technology, 2nd Edition, 3 Volumes. Chapman & Hall, London, 1997.

B5 Brockmann, W.; Geiss, P.L.; Klingen, J.; Schröder, W.: Adhesive Bonding – Materials, Applications and Technology. Wiley-VCH, Weinheim, 2009 (ISBN 978-3-527-31898-8).

B6 Burchardt, B.; Diggelmann, K.; Koch, St.; Lanzendörfer, B.: Elastic Bonding. Verlag moderne industrie, Landsberg 1998 (ISBN 3-478-93192-4).

C1 Chan, C.M.: Polymer Surface Modification and Characterization, Hanser-Verlag, München, 1994 (ISBN 1-56990-158-9).

C2 Cognard, Ph. (Ed.): Adhesives and Sealants – Basic Concepts and High-Tech Bonding. Vol. 1, Elsevier, New York, 2005 (ISBN 0-08-044544-3).

C3 Comyn, J.: Adhesion Science. Royal Society of Chemistry, 1997 (ISBN 0-85404-543-0).

D1 DeLollis, N. J.: Adhesives, Adherends, Adhesion. Robert E. Krieger Publishing Comp. Inc., Huntington, New York, 1980.

Applied Adhesive Bonding: A Practical Guide for Flawless Results. Gerd Habenicht
Copyright © 2009 WILEY-VCH Verlag GmbH & Co. KGaA, Weinheim
ISBN: 978-3-527-32014-1

12 References

F1 Flick, E. W.: Adhesives and Sealant Compound Formulations, 2nd Ed. Noyes Publications, Park Ridge, New Jersey, 1984.

F2 Flick, E. W.: Construction and Structural Adhesives and Sealants. An Industrial Guide. Noyes Publications, Park Ridge, New Jersey, USA, 1988.

G1 Gruber, W.: High-Tech Industrial Adhesives. Verlag moderne industrie, Landsberg, 2001.

H1 Hartshorn, S. R.: Structural Adhesives – Chemistry and Technology. Plenum Press, New York, London, 1986.

I1 Inagaki, N.: Plasma Surface Modification and Plasma Polymerisation. Technomic Publishing, Basel, 1996 (ISBN 1-56676-377-1).

I2 Israelachvili, J.: Intermolecular and Surface Forces, 2nd Edition. Academic Press, London, San Diego, 1992.

K1 Kinloch, A. J.: Adhesion and Adhesives. Science and Technology. Chapman and Hall, London, New York, 1987.

K2 Kinloch, A. J.: Durability of Structural Adhesives. London, Appl. Sci. Publ., 1983.

K3 Kinloch, A. J.; Young, R. J.: Fracture Behaviour of Polymers. Elsevier Applied Science Publishers, London, New York, 1985.

K4 Krevelen v., D. W.: Properties of Polymers, 3rd Edition. Elsevier, Amsterdam, New York 1990.

L1 Landrock, A. H.: Adhesives Technology Handbook. Noyes Publications, New Jersey, USA, 1985.

L2 Lee, L. H. (Ed.): Fundamentals of Adhesion. Plenum Press, New York, London, 1991.

L3 Lees, W. A.: Adhesives in Engineering Design. The Design Council, London, and Springer Verlag, Berlin, 1984.

M1 May, C. A. (Ed.): Epoxi-Resins-Chemistry and Technology, 2nd Edition. Marcel Dekker Inc., New York, Basel, 1988.

M2 Mittal, K. L.: Adhesive Joints. Formation, Characteristics and Testing. Plenum Press, New York, 1982.

M3 Mittal, K. L.; Pizzi, A.: Adhesion Promotion Techniques, Technological Applications. Marcel Dekker Inc., 1999, 404 p. (ISBN 0-8247-0239-1).

M4 Mittal, K. L.; Lee, K. W.: Polymer Surfaces and Interfaces: Characterization, Modification and Application. VSP-Verlag, Zeist, Netherlands, 1996 (ISBN 90-6764-217-7).

M5 Mittal, K. L.: Polymer Surface Modification: Relevance to Adhesion. VPS-Verlag, Zeist, Netherlands, 1995.

P1 Packham, D. E.: Handbook of Adhesion. Wiley-VCH, Weinheim, 2005 (ISBN 0-471-80874-1).

P2 Pizzi, A.: Wood Adhesives – Chemistry and Technology. Marcel Dekker Inc., New York, Basel, 1985.

P3 Pocius, A. V.: Adhesion and Adhesives Technology: An Introduction. Hanser Verlag, München, 1997 (ISBN 1-56990-212-7 and 3-446-17616-0).

P4 Possart, W. (Ed.): Adhesion – Current Research and Applications. Wiley-VCH, Weinheim, 2005 (ISBN 3-527-31263-3).

- **S1** Sadek, M. M.: Industrial Applications of Adhesive Bonding. Elsevier Applied Science; London, New York, 1987.
- **S2** Satas, D. (Ed): Handbook of Pressure-Sensitive Adhesive Technology. Satas and Associates, Warwick, R. I. 1999 (ISBN 0-9637993-3-9).
- **S3** Shields, J.: Adhesives Handbook, 3rd Edition. Butterworths, London, 1984.
- **T1** Thrall, E. W.; Shannon, R. W.: Adhesive Bonding of Aluminium Alloys. Marcel Dekker Inc., New York, Basel, 1985.
- **W1** Wake, C. W.: Adhesion and the Formulation of Adhesives, 2nd Edition. Appl. Sci. Publ., London, 1982.
- **W1** Wegman, R. F.: Surface Preparation Techniques for Adhesive Bonding. Noyes Publications, Park Ridge, USA, NJ, 1989.
- **W2** Wool, R. P.: Polymer Interfaces – Structure and Strength. Hanser Verlag, München, New York, 1995 (ISBN 3-446-16140-6).

13
A Selection of Common Terms in Bonding Technology

Note: Please search the chapters/sections index and the respective subject for technical terms not mentioned in the following list.

Abhesives: Coatings based on special silicone compounds with adhesive layer resistant properties on, for example, release papers. Application as substrate for adhesive labels, double-sided adhesive tapes, and so on.

Absorption:
1. Absorption of liquid adhesives in a porous, nonclosed surface. Not to be confused with **ad**sorption.
2. Absorption of radiation, for example, UV rays through glass, plastics (e.g., Plexiglas) that results in a delay in curing of radiation-curing adhesives.

Accelerator: Adhesive component reducing curing time.

Acrylic adhesive: A polymerization adhesive deriving from acrylic acid.

Activators: Chemical compounds able to trigger chemical reactions not possible without such compounds (e.g., activators as means of pretreatment of hard-to-bond plastic surfaces with anaerobic adhesives). In contrast to catalysts, activators participate directly in chemical reactions.

Active surface: Surface disposing of reactive properties (e.g., dipoles) due to mechanical, chemical or physical pretreatment or a coating (primer, activator).

Adherend: Material to be joined or already joined with another adherend.

Adherend fracture: Failure of a bonded joint under mechanical stress in the adherend material, thus, outside the adhesive layer. Indicates that the bond strength is higher than the adherend strength.

Adherend surface: Surface of an adherend to be bonded or already bonded.

Applied Adhesive Bonding: A Practical Guide for Flawless Results. Gerd Habenicht
Copyright © 2009 WILEY-VCH Verlag GmbH & Co. KGaA, Weinheim
ISBN: 978-3-527-32014-1

Adhesion: The adhesion of two different materials caused by atomic or molecular adhesive forces.

Adhesion promoter: Chemical compounds enhancing adhesive strength and/or ageing resistance of bonded joints in the form of adhesive additions or surface coatings (q.v. primers).

Adhesive: Nonmetal (working) material joining two or more adherends after being transferred from the liquid state into a solid adhesive layer.

Adhesive films: Consisting of two-component reactive adhesives applied to a nonadhesive substrate for transport and storage purposes that is removed prior to processing. Curing occurs via chemical reaction under application of heat and pressure. Adhesive films are also commercially available as physically setting films, see heat-sealing adhesive.

Adhesive force: Act between the adherend surfaces and the adhesive layer molecules and are mainly based on electrical interactions (dipoles).

Adhesive fracture: Failure of a bonded joined due to fracture in the boundary layer area of adherend and adhesive layer.

Adhesive layer: Cured adhesive between two adherends.

Adhesive strips: Consist of paper or kraft paper strips, if required reinforced, coated with an adhesive layer that can be activated by water or heat.

Adhesive tapes: Consisting of plastic, plastic foam, metal, paper or textile tapes with or without reinforcement, coated with a pressure-sensitive adhesive layer on one or two sides.

Adsorption: Addition of solid, liquid or gaseous substances to a surface.

Ageing: Modification of properties of bonded joints due to mechanical, physical and chemical influences generally resulting in strength reduction.

Amines: Chemical compounds with nitrogen as central atom, *inter alia* applied as hardener for epoxy resin adhesives.

Amorphous: Opposite of crystalline, characteristic for the structure of substances without crystalline portions.

Animal glue: Glue made of protein-containing products, particularly of animal waste (bones), on aqueous basis.

Anisotropic: Material property with behavior differing in dependence on stress direction, for example, wood, electrical properties of special conductive adhesives; opposite: isotropic.

Assembly time (closed): Period of time during which a bonded joint has to be fixed until it is strong enough to resist a displacement of the adherends by external force effects.

Assembly time (open): Period of time between the adhesive application and the fixing of the adherends.

Atmospheric pressure plasma: Physical-chemical procedure for surface pretreatment, in particular of plastics. It is based on the development of an ionized gas atmosphere by high voltage, and leads to the formation of active surfaces. In contrast to the low-pressure plasma, it works at atmospheric pressure.

Atoms: Smallest "components" of elements able to combine to molecules.

Autoclaves: Device in which the contemporaneous generation of high pressures and high temperatures is possible. Applied in adhesive processing mainly for polycondensation curing adhesives.

Batch: See Mix.

Binder: Adhesive component that, depending on kind and quantity, decisively determines the properties of the adhesive layer; also called raw material.

Blocked reactive adhesives: Adhesives in which the inter-reaction of the components is eliminated by mechanical (e.g., separated packing units, microencapsulation) or chemical measures (special formulations of resins, respectively, hardeners).

Bond strength: Force in relation to a defined area, required to separate a bonded joint.

Bonding: Joining of same or different materials applying adhesive.

Breaking stress: The stress required for breaking the material or a bonded joint.

Casein adhesive glue: Adhesive on animal basis, made of acid casein accruing from milk processing. Application as wood adhesive and in the packaging industry (label adhesive).

Catalyst: Chemical compounds able to trigger a chemical reaction that would not be possible without such compounds. In contrast to activators, catalysts do not take part in chemical reactions.

Cleaner: Cleansing agent for surfaces.

Cohesion: Inner strength of a material. In bonding technology, a term generally referring to the adhesive layer strength.

Cohesion fracture: Failure of a bonded joint due to fracture in the adhesive layer.

Cohesive forces: Act between molecules in the adhesive layer. Sufficient cohesive strength depends on the observance of the determined curing time and temperature as well as on a homogeneous mixture of the adhesive components.

Cold curing: Curing of adhesives without heat supply. Temperatures below room temperature delay reaction, while those above room temperature accelerate reaction.

Collagen: Protein product extracted from animal skin and bones. Basic material for gelatine glues.

Components: Elements of a reactive adhesive to be mixed prior to processing according to the mixing ratio determined by the adhesive manufacturer to achieve an equal and complete curing of the adhesive layer.

Conductive adhesives: Adhesives with adhesive layers able to conduct electric current (silver particles) or heat (aluminum oxide, boron nitride) due to the addition of respective fillers.

Contact adhesives: Adhesives solidifying to an adhesive layer after the evaporation of the solvent ("contact dryness") by pressure application.

Contact angle: Characterizes the wetting behavior of an adhesive (a liquid) on a surface. For good wetting, the value of the contact angle α should be below 30°.

Copolymer: Polymer consisting of two or more monomer units with different basic structures involved in polymerization.

Corona process: Method for surface pretreatment of plastics, based on the entrapment of reactive atoms from the gas phase in the surface of plastics by high-voltage discharge.

Corrosion: Damaging or modifying of (usually) metal materials in the surface area by chemical or electro-chemical reactions.

Creep: Permanent deformation of a bonded joint or a material after mechanical stress. In bonding technology, important for adhesive layers.

Crosslinking: Reaction resulting in the spacious linkage of molecule chains. The linkage is a kind of reaction when curing adhesives.

Crystallinity: In contrast to amorphous, the existence of crystalline parts in polymer structures (e.g., polyethylene, polyamides).

Curing: See Setting.

Curing conditions: Parameters essential for adhesive curing, for example, temperature, time, air humidity, and so on.

Curing time: see Setting time.

Cyanoacrylate adhesive: Quickly, within seconds curing reactive adhesives, with curing being initiated by moisture.

Density ratio (ρ): Ratio of mass m and volume V of a matter: $\rho = \dfrac{m}{V}$.

Units: $\dfrac{kg}{m^3}$; $\dfrac{kg}{dm^3}$; $\dfrac{g}{cm^3}$.

Dew point: Temperature value reached when a vapor-gas mixture reaches saturation and condensation occurs during further cooling, for example, condensation of water vapor, q.v. relative humidity (Section 4.2.2).

Dextrin: Chemically modified starch, raw material for dextrin glue.

Dextrin adhesive: Aqueous adhesive based on catabolized starch.

Diffusion: Independent mixing of gases, liquids, solids in and with each other due to molecular movement. Important for bonding, for example, diffusion of water vapor in adhesive/boundary layers or diffusion of solvents through porous adherends during the setting of adhesives.

Diluents: Liquid, organic compounds (combustible!) reducing solid concentration and/or viscosity of an adhesive.

Dipole: Molecule with different electric charge distributions.

Dispersion: Matter extremely finely and permanently dispersed in a liquid (e.g., a polymer).

Dispersion adhesive: Contains water as solvent in which polymer particles "float" owing to their extremely small particle size. After removal of the water, they set to an adhesive layer by "fusion" of the particles.

Dissociation: Decomposition of molecules in aqueous solvent in positively or negatively charged ions (q.v. ions).

Double bond: In organic chemistry, the combination of two carbon atoms by two valencies (C=C). The double bonds are the prerequisite for curing of polymerization adhesives.

Drying time: Period of time to pass when using cleansing agents, activators or primers before the solvent is completely evaporated and the adhesive is ready for application.

Dynamic mixer: Mixer allowing the mixing of different matters by means of mechanical energy, usually by rotary motions.

Elasticity: Property of a material to deform under the impact of a force and to regain the original state after load relieving.

Elastomers: In contrast to thermoplastics and thermoset material, loosely cross-linked polymer molecules that do not melt at increasing temperature and show high reversible extensibility.

Electromagnetic spectrum: Depiction of electromagnetic radiation according to radiation energy and wavelength. Important ranges (by decreasing wavelength and thus increasing energy): radio and microwaves, infrared/thermal radiation, visible light, UV, X-rays, gamma rays.

Elongation at break: Parameter indicating the elongation in relation to the original length (in %) of a material up to its fracture. In the case of elastomers also called ultimate elongation.

Emulsifier: Component of dispersions with the special property to hold polymer particles in the aqueous phase in suspension, thus preventing their sedimentation.

Emulsion: Liquid with fine dispersion of a second liquid. No solution.

Etching: Surface modification by removal of near-surface layers by means of acids (acid etching) or alkali solutions (alkaline etching).

Expansion coefficient: Parameter for the description of temperature-dependent changes of dimensions of a material or component.

Filler: Adhesive component in a solid, finely dispersed form that specifically modifies the processing properties of the adhesive and the properties of the adhesive layer (e.g., metal particles in electrically conductive adhesives, chalkstone, carbon black to increase viscosity). Fillers are not reactive partners in adhesive curing.

Final strength: Maximum achievable strength of an adhesive under standard conditions. The bond strengths (e.g., adhesive strength) as well as the material parameters of the adhesive (e.g., E-module, tear strength and elongation at break) indicated in adhesive data sheets are determined according to the final strength of the adhesive, that is, the completely cured adhesive.

Fixing: The fixation of the adherends, with or without pressure, in the desired position during the curing process.

Fixing adhesive: Adhesive used for the fixing of components prior to further processing, for example, components on printed circuit boards before soldering, wire ends for coils after winding. The adhesive layer is usually not exposed to mechanical stress.

Flame treatment: Surface pretreatment method, especially for plastics, by means of an acetylene, propane or butane flame burning in excess oxygen. Results in improved surface wettability by the adhesive due to the chemical entrapment of oxygen atoms in the polymer surface.

Flash point: Criterion for flammability of combustible liquids with subdivision into hazard categories. The flash point is the lowest temperature at which vapor develops from a liquid to such an extent that an inflamable vapor-air-mixture results, for example, hazard categories for adhesives or solvents:
A I   = flash point < 21 °C
A II = flash point 21–55 °C
A III = flash point > 55–100 °C

Flexural strength: The strength of a material in bending, expressed as the stress on the outermost fibers of a bent test specimen, at the instant of failure. Dimension N/mm^2 (MPa).

Fracture energy: The energy required for a material or a bonded joint to fracture.

Gel: Semisolid, colloidal system consisting of a solid dispersing in a liquid. Can be transferred into a sol state by increase in temperature or dehydration.

Gelatine: Protein product soluble, respectively, swellable in water, extracted from collagen.

Gel time: In two-component systems, the period of time in which a ready-to-use adhesive mix passes from the free-flowing to the sag-resistant state.

Glass transition temperature: (abbreviated T_g) A temperature or temperature range characteristic for polymers, below which they are in a hard/brittle state. Usually connected with a radical change of mechanical and physical properties of the polymers. Thermoplastics pass to the flow, respectively, melting range via plastic range.

Glue: Adhesive consisting of animal, plant or synthetic raw material and water as solvent.

Glueline: Space between two adherends to be filled with adhesive.

Glutin-adhesive: Water based adhesives, formulated from animal products, such as bones, skin.

Handling strength: Strength of a bonded joint that allows for further processing in the work flow.

Hardener: Adhesive component, triggering the chemical setting of an adhesive by polymerization, polycondensation or polyaddition, added or admixed to the adhesive resin, often also called "second component".

Heat-curing adhesive: Adhesive requiring a predetermined temperature–time profile for curing.

Heat-sealing adhesive: A hot-melt adhesive applied to the adherends as solid adhesive layer, which melts under heat/pressure and joins the adherends after cooling.

Homogenization: Equal dispersion of two or more substances in liquid/liquid or liquid/solid systems obtained by mixing or stirring.

Homopolymer: Polymer consisting of only one kind of monomer element with similar kind of monomer reaction.

Hot-melt: Adhesive applied as melt at elevated temperature, physically setting when cooled, see hot-melt adhesive.

Hot-melt adhesives: Hot-melt adhesives are applied to the adherends in a fused form and set by cooling. Their "open assembly time" is very short, thus adherends have to be fixed immediately after adhesive application.

Inhibitor: Substance that, in low concentration, inhibits chemical reactions, for example, oxygen inhibitation of radiation-curing adhesives.

Initial strength: Strength, also handling strength, an adhesive develops shortly after the application and fixation of the adherends; adhesive property important for further processing.

Initiator: Substance triggering a chemical reaction already in low concentration, for example, photoinitiators in UV-radiation-curing adhesives.

Ions: Positively or negatively charged atoms or molecules, for example, Na^+ and Cl^- ions in aqueous saline solution (q.v. dissociation).

Isotropic: (Iso (Greek) = same; tropos (Greek) = direction) Same properties of a substance in all directions (e.g., electric properties of special conductive adhesives) Opposite: anisotropic.

Joining: The putting together of adherends. In bonding technology with ensuing transition of the adhesive into the adhesive layer.

Kelvin scale: Unit K; temperature scale, with the reference point 0 K corresponding to the absolute zero point −273.16 °C; thus 0 °C = 273,16 K, 100 °C = 373.16 K, and so on. The difference from the Celsius scale is the fact that the Kelvin scale does not refer to properties of special substance. A temperature difference of for example, −20 to +40 °C is thus indicated with 60 K, the boiling point of water as a substance-constant with 100 °C.

Laminating: Joining of mostly large area, flexible adherends (e.g., films, veneers) to a composite material by means of adhesives.

Light-curing adhesives: In contrast to UV-curing, adhesives curing under exposure to visible light in the wavelength range of 400–500 nm (nanometers).

MAK-value: Maximum workplace concentration, product-specific value of chemical substances defining the harmful contamination of the air at the workplace due to these substances (dimension: ppm = parts per million = mg/kg). Information on MAK-values are given in the safety data sheets of the respective substances.

Maximum drying time: Period of time after the application of the adhesive just allowing for bonding. If the maximum drying time is exceeded, the polymer layers on the adherends will already have solidified to such an extent that they are no longer capable of developing a strong adhesive layer.

Mechanical adhesion: Development of adhesive forces by positive joining of the adhesive layer in geometrical surface structures (pores, capillaries, roughnesses).

Methyl cellulose: An ether generated by the partly methylation of the hydroxyl groups of cellulose.

Microencapsulated adhesive: Reactive adhesive mixture, with the (liquid) components encapsulated by a protective skin in the form of finest drops, preventing a reaction during storage. Only after the destruction of the capsule wall, for example, by screwing a nut onto a suchlike coated screw, does a chemical reaction set in and an adhesive layer develop.

Minimum drying time: In the case of solvent-based adhesives, the time period between the adhesive application and the fixing of the adherends, to enable the evaporation of the major part of the solvent from the liquid adhesive film.

Mix: Adhesive composed for application according to a determined mixing ratio of the individual components.

Mixed fracture: Failure of a bonded joint by proportionate forms of adhesion and cohesion fracture, usually caused by improper adhesive processing and surface pretreatment.

Mixed glue: Combination of animal and/or plant glues with synthetic adhesives.

Mixing ratio: Ratio for the mixing of the adhesive components prior to processing, prescribed by the adhesive manufacturer. Very important to achieve maximum bond strength.

Modulus of elasticity: Material parameter describing the relation of tension to expansion, when a material is under mechanical stress. Dimension N/mm^2 (MPa).

Molecules: Chemical substances, consisting of same or different atoms.

Monomer: Base products of an adhesive, developing polymer molecule structures (adhesive layers) due to chemical reactions.

Nanometers: Unit length, 1 nm = 10^{-9} m = 0,000 000 001 m.

Natural adhesives: Adhesives, mainly made of natural products (protein, latex, starch).

Natural rubber: Wide-meshed crosslinked polymer with rubber-like properties made of plant milk (rubber tree).

Oligomers: Polymers with an only limited number of monomers (e.g., dimers – two monomers; trimers – three monomers).

One-component adhesive: Adhesive that does not require mixing with another adhesive component prior to application.

Open assembly time: Maximum period of time between the application of the adhesive and the fixing of the adherends.

Oxides: Chemical compounds of elements with oxygen, for example, iron oxide (FeO/Fe_2O_3 = rust), but also water (H_2O = hydrogen oxide) or carbon dioxide (CO_2).

Pascal: Dimension for mechanical stress and pressure; in bonding technology for example, dimension for bond strength (MPa = 10^6 Pa).

Paste: Adhesive in the form of an aqueous swelling product that, in contrast to glues, develops a high-viscosity nonropey mass already in low raw material concentrations.

Peel resistance/strength: Resistance of a bonded joint against linear acting peel forces generating high stress peaks in the adhesive layer, dimension N/mm or N/cm.

Photoinitiator: Substances triggering the polymerization reaction for UV or light radiation-curing adhesives.

Physically setting adhesives: Adhesives already existing in the form of polymers transferred into a liquid form by solvents or water, respectively, by melting, forming an adhesive layer after evaporation or cooling (e.g., solvent-based, dispersion and hot-melt adhesives).

Pickling: Chemical removal of reactive layers on metal materials by means of diluted acids.

Plasma: Gas state (also called fourth dimension) from a mixture of electrically charged ions and neutral atoms/molecules. Results from plasma discharge; mainly used for surface pretreatment of plastics (q.v. atmospheric plasma).

Plasticizer: Organic compounds with low molecular weight physically integrated in the polymer structure, that is, not by a chemical reaction, thus contributing to higher deformability and/or plastification of the polymer. Under adequate conditions (higher temperatures) they can diffuse out of the polymer structure (plasticizer migration).

Plasticizing: See Plasticizer.

Plastisol: Adhesive consisting of polymers (e.g., polyvinyl chloride) in plasticizers, solidifying to an adhesive or sealing layer by (physical) entrapment of the polymer in the plasticizers when heated (sol-gel transformation).

Polarity: Differences in electric charges in molecules. In adhesive molecules in particular responsible for the development of adhesive forces. See dipole.

Polyaddition: A chemical curing reaction in which two differently structured monomers A and B accumulate to form a polymer AB.

Polycondensation: In contrast to polyaddition and polymerization adhesives, a byproduct, for example, water, develops during curing. Apart from the application of heat, adequate pressurizing of the adherends is required (see autoclave).

Polymer: Chemical compound consisting of monomers or prepolymers developed by polyaddition, polycondensation or polymerization, usually in solid state. In cured condition, adhesives generally consist of polymers.

Polymerization: Formation of polymers from monomers or prepolymers having a C=C double bond, for example, acrylate adhesives.

Pot life: Period of time after mixing of the components in which a reactive adhesive needs to be processed.

Prepolymer: Preliminary stage of polymers with still existing reactive properties.

Pressure-sensitive adhesive: Adhesive present in solvents or available as dispersion, which remains permanently adhesive after setting.

Pretreatment: See Surface pretreatment.

Primers: Substances enhancing adhesion between adherend surfaces and adhesive and delaying ageing processes. In contrast to adhesion promoters, primers are applied to the adherend surfaces. Adhesion promoters are usually admixed to the adhesive (q.v. adhesion promoter).

Processing temperature: Temperature of the adhesive or its natural surroundings during processing.

Processing time: Maximum available period of time for processing after completed mixing of a multicomponent adhesive.

Radiation-curing adhesives: Adhesives cured by electro-magnetic radiation, in particular by UV radiation or by visible light.

Radiation-source: In bonding technology, equipment for the curing of UV-electron beam and light-curing adhesives.

Raw material: See Binder.

Reaction rate: Periodic change of concentrations in the case of chemical reactions.

Reactive adhesive: Adhesive curing due to a chemical reaction with curing depending on time, temperature, pressure or humidity.

Reactive group: Chemical "linkage points" on monomers or prepolymers enabling the formation of polymers via chemical reactions, for example, epoxy or amine groups in the case of epoxy resin adhesives.

Reactive hot-melt adhesives: Adhesives setting both by cooling of the melt and also by an ensuing chemical reaction.

Reduction factors: Factors considering ageing and production-related influences in strength calculations.

Relative humidity: See Section 4.2.2.

Release paper: Papers abhesively coated with special silicones to wind up adhesive tapes or to provide the application-specific availability of adhesive labels, see Abhesives.

Resin: Collective term for solid and viscous, organic, noncrystalline products with more or less broad dispersion of the molar mass. Usually, resins have a melting or softening range, are brittle in the solid state and break in a clam-like manner. They tend to flow at room temperature. Apart from resins as additives to adhesives, some adhesive raw materials, for example, epoxy resins, phenol resins, polyester resins bear this name, too.

Rheology: Subarea of physics dealing with the description, explanation and measuring of the flow behavior of substances capable of flowing.

Rubber: Wide-meshed crosslinked polymers with low glass transition temperatures. Differentiation into natural (natural rubber) and synthetic (nitrile, butyl, chloroprene, styrene) rubbers.

Safety data sheet: Form to be compiled by the manufacturer regarding special properties of chemical substances, in particular regarding their possible sources of danger.

Sedimentation: Sedimentation of fillers in liquid adhesives.

Setting: Solidification of adhesives through physical and/or chemical processes. In the case of setting through chemical processes, solidification occurs by means of molecule enlargement and crosslinkage of monomers and/or prepolymers.

Setting time: Period of time in which the adhesive, after joining, reaches the degree of crosslinkage required for the application-related stress.

Shear modulus: Ratio of shear stress and shearing strain in the case of simple shear deformation.

Silanes: Organic silicon compounds, in particular used to improve adhesive and ageing properties of surfaces.

Silicones: Adhesives and sealing compounds with the basic framework based on –Si–O– bonds. Available as one and two-component systems (RTV-1 and RTV-2). They are characterized by high-temperature resistance as well as high ageing resistance.

Skinning: Beginning superficial solidification (curing) of a reactive adhesive applied to an adherend. Skinning impedes or even prevents the wetting of the second adherend. Examples: polyurethane, silicone adhesives.

Skinning time: Period of time from adhesive application to the beginning superficial solidification of the adhesive, when bonding is no longer possible.

Softening point: Temperature or (usually) temperature range, characterizes the transition from solid to plastic/paste-like and then liquid state of a material, for example, glasses, thermoplastics (q.v. glass transition temperature).

Sol: See Gel.

Solvent: Organic liquids capable of solving other substances (e.g., polymers) without changing themselves or the solved substance. Application for example, as cleaning agent, dilutor or volatile components in solvent-containing adhesives.

Solvent-based adhesives: Adhesives with adhesive-layer forming substances (polymers) being solved or pasted. Depending on the character of the adherends, the solvents must evaporate completely or partly prior to fixing.

Spreading: Unimpeded spreading of a liquid on a surface, represents the optimum way of wetting.

Stabilizer: Adhesive component serving to maintain the property of the adhesive and/or of the adhesive layer regarding storage, processing and stress impacts.

Starch: Plant product, so-called carbohydrate, consisting of the elements carbon, hydrogen and oxygen. Raw material for aqueous adhesives (paste, glues).

Starch adhesive: Aqueous adhesive based on natural starch.

Static mixing tube: Mixing device for two-component adhesives, mainly used for adhesives with identical mixing portions and processing viscosities. Mixing occurs by layer formation via staggered mixing impellers.

Storage stability: Period of time in which a substance, for example, an adhesive, stored under specified conditions retains its application properties; must not be exceeded prior to adhesive application.

Strength: Resistance of a material against a deformation caused by forces acting on it from outside. Dimension: N/mm^2 (MPa), q.v. bond strength.

Surface tension: Measure for "internal strength" of liquids characterizing their wetting behavior of surfaces. Liquids with low surface tension disperse equally on surfaces, high surface tensions result in rolling off from the surface (e.g., mercury drops).

Surface treatment: Generic term for processes applied to surfaces to achieve a surface condition of the adherends suitable for a bonded joint or to optimize them in view of their adhesiveness. Such processes are divided into surface preparation, pretreatment and post-treatment.

Synthetic adhesives: Adhesives based on raw materials usually not of natural origin.

Temperature resistance: An important adhesive property for the application at elevated operating temperatures. If this temperature is exceeded, chemical decomposition will set in and with this an irreversible damaging of the adhesive layer.

Tensile shear strength: Strength of a single-lap bonded joint by an eccentrically attacking force until it breaks (Figure 10.2).

Tensile strength: Tensile stress at break of a material, respectively, a bonded joint under tensile stress.

Thermal conductivity: Ability of a material to conduct or transfer heat. Dimension: W/cm K (Watt per centimeter Kelvin). For values of selected materials, see Section 9.1.1.4.

Thermal expansion: Temperature-dependent volume, respectively, longitudinal expansion of a body. Measure is the thermal expansion coefficient α in the dimension 10^{-6} K^{-1}; Values α: steels 10–20, aluminum and Al-alloys 20–25, glasses 5–10, plastics/adhesive layers 50–100.

Thermal stability: Temperature and time behavior of a bonded joint that do not provoke changes of strength characteristics.

Thermoplastics: Plastics (adhesive layer) with predominantly straight or branched polymer structures that pass from solid via soft/plastic to the fused state. Certain kinds of thermoplastics are soluble in organic solvents.

Thermoset material: Plastic/adhesive layer consisting of molecule structures closely crosslinked by covalent bonds. A thermoset material is neither fusible, nor plastically deformable and is insoluble in solvents.

Thixotropy: Property of certain liquid systems to temporarily adopt lower viscosity after adding so-called thixotropic agents (e.g., silica products) under mechanical influence (e.g., stirring, coating). For adhesives, this bears the following advantages: no running-off on vertical surfaces, achievement of higher adhesive layer thicknesses, avoidance or reduction of adhesive penetration into porous adherend surfaces.

Thoughened adhesives: Adhesives with rubber-elastic components chemically integrated in their polymer network to enhance their mechanical properties.

Two-component adhesives: Chemically reacting adhesives with a second component (hardener) to be admixed to another component (usually the resin component).

Type of adhesive: Adhesives based on different adhesive raw materials with special processing properties (e.g., hot-melt adhesives, pressure-sensitive adhesives), different purposes of use (e.g., wallpaper paste, wood glue), processing temperatures (e.g., cold glue, heat-curing adhesives), availabilities (e.g., adhesive films, solvent-based adhesive).

UV curing: Curing of adhesives by means of electromagnetic radiation at wavelengths in the range of UV-A 315–380 nm, UV-B 280–315 nm, UV-C 200–280 nm (nm = nanometers). Radiation energy increases with decreasing wavelength.

Valencies: Adhesive forces between congeneric or different atoms as basis for the development of molecules and polymers, respectively.

Viscosity: Parameter describing the flow properties of a liquid depending on the internal friction of the molecule. It is defined by the force in Newton (N) required to displace one interface parallel to the opposite interface with a speed of 1 m s^{-1} in a liquid layer with a surface area of 1 cm^2 and a height of 1 cm. Unit: (Pa s), Pascal second. In bonding technology, the dimension mPa s = 10^{-3} Pa s, milli-Pascal second is often applied for calculation. Water, for example, has a viscosity of 1 mPa s.

Wavelength: Denomination λ, measure for the length of a periodical oscillation of electromagnetic waves (see electromagnetic spectrum), unit nm (nanometer).

Wetting: The ability of liquids to spread equally on solid matters. In bonding technology, the property of an adhesive to spread equally on the surfaces of the adherends. The wettability of a system depends on the respective surface tension of the solid and liquid media.

Index

a
abhesives 149
absorption 52, 149
AB-method (methacrylates) 36
ABS (acrylonitrile-butadiene-styrene-copolymers) 114
accelerator 149
acetone 65
acrylonitrile-butadiene-styrene 110
acrylic glass 114
activator 149
active surface 149
adherend 4, 5, 7, 149
– combinations 103
– fracture 149
– properties 95
– surface 3, 4, 58, 66, 149
adhesion 57, 62, 63, 102, 150
– fracture 134, 150
– promoter 150
adhesive 1, 3, 4, 9, 11, 19, 150
– all purpose 11
– classification 5, 8, 9
– chemical basis 9, 10
– chemically reactive 8, 9, 41
– cold curing 18, 75
– conductive 157
– heat activateable 55
– heat curing 75
– heat sealing 46
– inorganic based 10, 56
– natural based 10, 55
– one-component 17, 18
– organic based 10
– physically setting 8, 9
– polyaddition 24
– polycondenzation 39
– polymerization 32
– reactive 13
– solvent based 8
– solvent free 9
– structure 5
– synthetic basis 10
– two-component 14, 17, 18
– water activateable 55
adhesive amount (surface roughness) 77, 78
adhesive application 63, 75
adhesive bonder 91
adhesive bonding 1–3
– advantages 1
– disadvantages 1, 3
– mistake possibilities 85–87
– remedial actions 85–87
– training 91
adhesive curing 63, 79, 80
– pressure application 79
adhesive drying 79, 80
adhesive engineer 91
adhesive film 150
adhesive forces 57–59, 61, 63
adhesive layer 3, 4, 57, 58, 150
– properties 19
– surface roughness 45, 77, 78
– thickness 78
adhesive mixing 72
adhesive preparation 71
adhesive processing 71, 88, 89
– safety measures 88
– workplace prerequisites 88
adhesive selection 93, 94, 101
adhesive specialist 91
adhesive strength 126, 130
– testing 128
adhesive strips 55, 150
adhesive tapes 53, 100, 104, 150
– foamed 53, 54
– one-side 53, 54

Applied Adhesive Bonding: A Practical Guide for Flawless Results. Gerd Habenicht
Copyright © 2009 WILEY-VCH Verlag GmbH & Co. KGaA, Weinheim
ISBN: 978-3-527-32014-1

– permanent 54
– removable 54
– repositionable 54
– transfer 53, 54
– two-side 53, 54
adhesive types 23, 45
– acrylic resin 31, 149
– anaerobic 18, 37, 41, 99
– contact 50
– cyanoacrylate 18, 33, 98
– dispersion 51
– epoxy resin 18, 23, 25, 98
– hot melt 8, 45
– methacrylate 18, 35, 41, 99
– phenolic resin 39
– plastisol 53
– polyester resin 18
– polyurethane 18, 26, 31, 98, 99
– pressure sensitive 8, 53, 100, 104, 150
– radiation curing 34, 41, 99
– silicones 40, 98, 99
adjusting (adherend) 64
adsorption 150
advantages, bonding 1
aerospace industry 68
ageing 150
all purpose adhesive 11
aluminum 107
amine 23, 150
amorphous 150
animal glue 55, 150
anisotropic 151
application, adhesives 75, 76
assembly time 45, 48, 151
atmospheric pressure plasma 113, 151
autoclave 151
autohesion 114

b
Baekeland, L.H. 6
bakelite 6
bending 1
benzene 65
binder 151
blasting 67, 68
blocked reactive adhesive 151
blocking (adhesives) 17, 19
bonded joint 3, 4, 63
– constructive design 139
– production 63
bonding 1–3, 151
– advantages 1
– disadvantages 1

– repair 81, 82
– structural 4
bond strength 15, 151
– values 129
booster 28
boundary layer 3, 4, 62
brass 108
breaking stress 151
brushing 67
budene 111
buna 111
butt joint 142
butyl rubber 111

c
carbon 5
carbon-carbon double bond 6, 31, 32
carbon reinforced plastic 111
casein glue 56, 151
catalyst 151
cellulose 56
cement mortar 43
ceramic 123
chemical element 5
chemical structure 5
– branchings 5
– chains 5
– crosslinked 5
chemically reactive adhesives 8, 9
chemistry
– inorganic 5
– organic 5
CIPG-method (sealing) 43
clamping 1, 2
classification (adhesives) 8
cleaner 151
cleaning 64
cleavage stress 139
climatic conditions 102
climatization, adherend 64, 71
– adhesive 72
climbing drum peel test 131
cohesion 61, 62, 125, 152
cohesion force 152
cohesion fracture 134, 152
cohesive strength 63, 125
cold curing adhesive 18, 75, 152
cold welding paste 114
collagen 152
compact sealing 43
components 13, 152
– mixing ratio 14
conductive adhesives 152
constructive design 139, 140, 143

contact adhesive 48, 50, 100, 152
- double sided bonding 50
- one sided bonding 50
- polymers 50
contact angle 59, 152
cooling time 152
copolymer 152
copper 107
corona process 112, 152
corrosion 153
- surface 69, 70
crash resistance 26
creep 22, 152
creep corrosion 69, 70
crosslinking 153
crystallinity 153
cured-in-place-gasking 43
curing 4, 78, 80
- impact of temperature 16, 17
- impact of time 15, 17
curing conditions 153
curing time 80, 103, 153
cyanoacrylate adhesives 18, 33, 41, 153
- application 33
- safety measures 33, 34

d
definitions 3
degreasing 64, 66, 68
   agents 65
density 153
destructive test methods 128
dew point 153
dextrin glue 55, 56, 153
diffusion 153
diffusion bonding 84, 114, 115
diluent 153
dipole 58, 153
dipole force 58
direct glazing 29
disadvantages, bonding 1, 3
dispersion 8, 9, 51, 100, 153
- polymers 52
- polyurethane 30
dissoziation 154
double bond 154
dry bonding (sealing) 43
drying, adhesives 79, 80
drying time 47, 154
- maximum 47
- minimum 47
dynamic mixer 73, 154

e
elastic bonding 135
elasticity 154
elastomer 21, 41, 42, 110, 111, 154
electric forces 58
electromagnetic spectrum 154
elongation at break 154
emulsifier 154
emulsion 154
epoxy group 23
epoxy resin adhesive 18, 23, 41, 98, 99
- application 26
- properties 26
epoxy resins 116
epoxy resin plastic 111
epoxy resin reactive hot-melt 25
etching 154
ethylacetate 65
ethylene-propylene-rubber 111
European Community 88
- standards (EN) 128
- welding federation 91
expansion coefficient 117, 120, 154

f
fiber-glass-reinforced plastic 111
filler (adhesive) 71, 155
film adhesive 42
film formation (dispersion) 52
final strength 155
finite element method (FEM) 131
FIPG-method (sealing) 43
fixing adhesive 155
fixing, adherends 78, 155
flame treatment 113, 155
flash point (solvents) 155
flexural strength 155
foam sealing 43
folding 1, 2
force 125, 126
- adhesive 57
- cohesive 62
- dipol 58
- intermolecular 58
formaldehyde 39
formed-in-place-gasket (FIPG) 43
fracture energy 155

g
galvanized steel 108
gel 53, 155
gel point 14
gel time 156
gelatine 155

Index

German Institute for Standardization (DIN)   128
German Welding Society (DVS)   91
glass   118
– radiation curing adhesives   119, 120
– surface pretreatment   118
glass/glass bonding   119
glass/metal bonding   120
glass transition temperature   20, 21, 156
glue   11, 56, 156
glueing   2, 40, 56
glueline   3, 4, 56, 156
glue stick   55
glutin adhesive   156
gold   107
grinding   67
grit blasting 67
gum arabic   56
gumming   55
guttapercha   111

h

handling strength   156
hardener   14, 23, 156
heat activateable adhesive   55
heat curing adhesive   75, 156
heat sealing adhesive   46, 156
heating time   80
hevea brasiliensis   121
homogenization   71, 156
homopolymer   156
hostaflon   110
hostalen   110
hot-melt adhesive   8, 9, 77, 100, 156
– advantages   45
– applications   46
– base materials   45
– reactive epoxy resin   25
– reactive polyurethanes   29
humidity   18
– relative   27
hycar   111
hydrogen   6
hydrogen-peroxide   35
hydroxide group   26
hydroxyl-polyurethan   30

i

indented joining   1, 2
inhibitor   157
initial strength   157
initiator   157
inorganic adhesives   10, 56
intermolecular forces   58

ions   157
isocyanate group   26
isopropylalcohol   65
isotropic   157

j

joining   1, 157
joint   1, 2
– nonpositive   1
– positive   1
– riveted   2
– screwed   2

k

Kelvin scale   157
Kreidl method   113

l

laminating   76, 77, 157
laminating resins   77
light-curing adhesives   157
liquid sealing   43
long-term stress   133
– test methods   133, 134
low-pressure plasma   113
lupolen   110
luting   2

m

macromolecule   19
MAK-values (max. workplace concentration)   89, 157
makrolon   110
materially joining process   1, 2
maximum drying time   48, 157
mechanical adhesion   57, 158
– interlocking   57
– surface pretreatment   67, 114
megapascal   127
melamine resins   111
melamine-formaldehyde adhesive   41
merlon   110
metals   102, 105, 107
– adhesive properties   105
– strength   105
– thermal conductivity   106
metal/plastic bonding   117
methacrylat adhesives   18, 35, 41, 99
– mix-system   36
– no-mix-system   36
– processing   35, 36
methylalcohol   65
methylcellulose   158
methyl-ethylketon   65

methyl-methacrylat   35
microencapsulated adhesive   158
minimum drying time   48, 158
mix   158
mixed fracture   134, 158
mixed glue   158
mixer, dynamic   73
– static   74
mixing, adhesives   72
mixing ratio   14, 15, 158
– bond strength   14, 15
mix-system (methacrylates)   35
moisture   102, 104
moisture-curing adhesives   27, 40
– one-component polyurethane   27, 28
– one-component silicones   40, 41
molecule   58, 158
monomer   6, 7, 13, 158
mortars   43
– hydraulically setting   43
– nonhydraulically setting   43
MS-polymers   42

n
nanometer   158
natural based adhesives   10, 55, 158
natural rubber   158
neoprene   111
Newton, J.   126
nitrile rubber   111
nitrogen   6
no-mix-system (methacrylate)   36
non-destructive test methods   128
nonpositively joint   2
novodur   110
novolen   110

o
oligomer   159
one-component adhesive   159
one-component epoxy resin adhesive   25
one-component polyurethan adhesive   27
one-component reactive adhesives   18, 98
open-assembly time   45, 48, 104, 159
organic based adhesives   10
oxides   159
oxygen   6
oxygen-oxygen-bond   35

p
paper, paperboard   123
Pascal, B.   127
Pascal, dimension   127, 159
paste   11, 56, 159

peel diagram   132
peel resistance   131, 159
– test method   131
peel strength   133, 159
peel stress   139, 140
perbunan   111
perchloroethylene   65
peroxide   35
petrolether   65
phase   51, 52
– gas   52
– liquid   52
– solid   52
phenolic adhesive   39, 41, 122
– plastic   111
– resins   116
physically setting adhesives   8, 9, 99, 159
photoinitiator   34, 159
pickling   69, 159
plasma   159
plastics   109
– classification   110
– identification   112
– surface pretreatment   112
plastic foams   116
plastic/metal bonding   117
plastisol   53, 100, 160
plastiziser   53, 118, 159
platinum   107
plexiglas   35, 110, 114
polarity   160
polyaddition   23, 41, 160
polyamide   110
polybutadiene   111
polycarbonate   110, 114
polychloroprene   111
polycondenzation   39, 41, 122, 160
polyester, unsaturated   39, 77
polyethylene   110, 116
polyethylene-terephthalate   110
polyisocyanate   27
polyisoprene   111
polymer   6, 7, 160
– formation   7, 13
– structure   7, 20
polymerization   32, 41, 160
polymer mortars   43
polymethylmethacrylate   35, 110
polyol   27
polypropylene   110, 116
polystyrene   110, 114
– adhesive   50
– foam   50
polysulfide   42

polytetrafluoroethylene   110
polyurethane adhesives   18, 26, 27, 41, 98, 135
– classification   31
– dispersion   30
– moisture curing   28
– one-component   18, 27, 31
– reactive hot-melt   29
– solvent based   30, 31
– solvent free   27
– two-component   27, 30, 31
polyurethan plastics   111, 116
polyvinylchloride   53, 84, 110, 114
porcelain   123
porous materials   103, 123
positive joint   1, 2, 57
post-treatment   70
pot life   14, 75, 103, 160
prepolymer   7, 13, 160
pressing   1, 2
pressure application (curing)   79
pressure-sensitive adhesives (PSA)   8, 9, 53, 160
– foamed   53, 54
– one-side   53, 54
– permanent   54
– removable   54
– repositionable   54
– structure   53, 54
– transfer   53, 54
– two-side   53, 54
pretreatment   64
– chemical   64, 68
– mechanical   64, 67
– physical   64, 68
primer   64, 70, 160
processing temperature   160
– time   160
production (bonded joints)   63

q
quality   81
– assurance   90
– management   91

r
radiation curing   34, 41, 104, 161
– UV   34, 104, 161
reaction heat   14
reaction layer   69
reaction rate   161
reactive adhesive   8, 13, 23, 41, 80, 161
reactive group   7, 13, 161
reactive hot-melt adhesive   25, 29, 161

reactivity   14
reduction factor   161
relative humidity   27
release liner/paper   54, 161
repair bonding   81, 82
– fiber fabric   84
– metal   81, 82
– plastics   83
– PVC-films   84
resin   14, 161
rheology   161
riveting   1
room-temperature-vulcanization (silicones)   40, 42
roughness, surface   67, 77
– adhesive amount   77
RTV 1,2 (silicones)   40, 41
rubber   121, 161
rubber/metal bonding   121

s
SACO-method (surface treatment)   68, 114
safety data sheet   88, 162
safety measures   88
sanding   67
screwing   1
sealing   2
sealing materials   42
– compact sealing   43
– dry bonding   43
– elastomeric   42
– foam sealing   43
– liquid sealing   43
– MS-polymers   42
– polysulfides   42
sedimentation   162
setting   4, 9, 162
– speed   103, 104
– time   162
shaft-to-hub joint   38, 137, 142
shear modulus   162
shear-strength   131
– test   131
shear testing method   131
short-time stresses   133
– test methods   133
silanes   162
silicone adhesives   10, 98, 111, 162
– one-component   18, 40, 98
– room-temperature-vulcanization   40
– two-component   41, 99
silicone release liner   54
silopren   111

silver 107
single-lap test 127
skinning 162
soap gel 65
sodium silicate 56
softening point 162
sol 53
sol-gel process 53
solder, soldering 1–3
solidification 9
solvent 162
solvent based adhesives 8, 9, 99, 162
– polyurethane 30, 31
– processing 48, 49
solvent free adhesives 9
– polyurethane 27, 31
spreading 59, 163
stabilizer 163
stainless steel 107
starch glue 55, 56, 163
static mixer 74, 163
steel 108
steel surface, blasted 67
storage stability 163
strength 125, 126, 163
stress distribution 2, 129, 130
structural bonding 4
structure, chemical 5
styrofoam 117
styron 110
surface contamination 66
surface layers 69
surface roughness 67
surface tension 60, 61, 163
– values 1
surface treatment 63, 64, 163
– metals 106
– plastics 112
– post-treatment 64, 70
– preparation 64
– pretreatment 64, 66, 68, 106
superglue 33
synthetic based adhesives 10, 163

t

teflon 110
temperature 15
– impact of curing 16
temperature resistance 163
tensile shear 141
tensile shear strength 128, 130, 163
tensile strength 163
tensile stress 141
tensile test 125

tension 129
terluran 110
terms 3, 4
tetrahydrofuran 84, 112
textiles 123
thermal conductivity 106, 164
– values 106
thermal expansion 164
– resistance 102, 106, 163
– stability 164
thermoplastics 19, 21, 41, 102, 110, 113, 164
– adhesives 45
– polymer structure 20
thermoset plastics 20, 21, 41, 103, 110, 111, 113, 164
– polymer structure 20
thixotropy 164
time 15
– cooling 80
– curing 80
– heating 80
– impact of curing 15
torsional moment 137, 138
toughened adhesives 164
T-peel-test 131
training (adhesive bonding) 91
treatment, see pretreatment
trichloroethylene 65
tubular bonded joint 142
two-component adhesives 14, 17, 18, 99, 164
– epoxy 23, 24, 99
– epoxy reactive 24
– polyurethane 27, 99
– polyurethane reactive 29

u

ultradur 110
ultramid 110
unsaturated polyester resins 39, 111
urea-formaldehyde adhesive/resin 41, 111
urethane-group 27
UV-radiation 34, 35, 119, 120, 164

v

valency 5, 24, 164
vapor degreasing 65
vestodur 110
vestalit 110
vestyron 110
vinnolit 110
viscosity 59, 60, 71, 103, 164
vulcanization 121

W

water activateable adhesive 55
waterglass 56
wavelength 165
welding 1–3
wet-adhesive 48
wet-bonding (contact adhesive) 50
– sealing 43
wettability 59
wetting 59, 60, 165
white glue 122
wood 40, 52, 122, 123

Z

zinc 108
zinc-plated steel 108